Lecture Notes in Mathematics

Edited by A. Dold and B. Eckmann

790

Warren Dicks

Groups, Trees and
Projective Modules

Springer-Verlag
Berlin Heidelberg New York 1980

Author
Warren Dicks
Department of Mathematics
Bedford College
London NW1 4NS
England

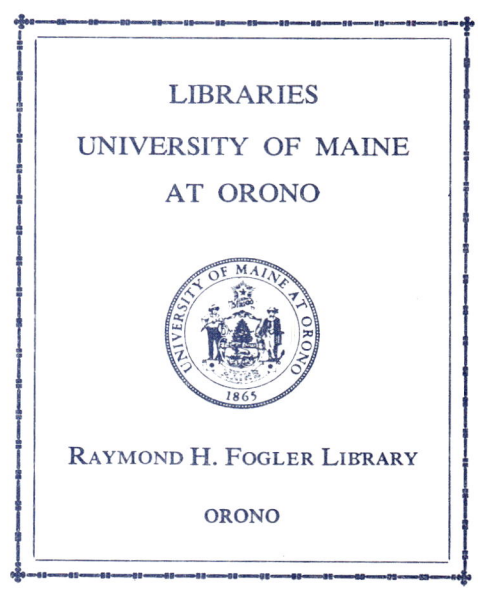

AMS Subject Classifications (1980): 16 A 50, 16 A 72, 20 E 06, 20 J 05

ISBN 3-540-09974-3 Springer-Verlag Berlin Heidelberg New York
ISBN 0-387-09974-3 Springer-Verlag New York Heidelberg Berlin

Library of Congress Cataloging in Publication Data. Dicks, Warren, 1947- Groups, trees, and projective modules. (Lecture notes in mathematics; 790) Bibliography: p. Includes indexes. 1. Associative rings. 2. Groups, Theory of. 3. Trees (Graph theory) 4. Projective modules (Algebra) I. Title. II. Series: Lectures notes in mathematics (Berlin); 790. QA3.L28 no. 790. [QA251.5]. 510s. [512'.4]. 80-13138

This work is subject to copyright. All rights are reserved, whether the whole or part of the material is concerned, specifically those of translation, reprinting, re-use of illustrations, broadcasting, reproduction by photocopying machine or similar means, and storage in data banks. Under § 54 of the German Copyright Law where copies are made for other than private use, a fee is payable to the publisher, the amount of the fee to be determined by agreement with the publisher.

© by Springer-Verlag Berlin Heidelberg 1980
Printed in Germany

Printing and binding: Beltz Offsetdruck, Hemsbach/Bergstr.
2141/3140-543210

*To the memory
of my mother*

PREFACE

For 1978/9 the Ring Theory Study Group at Bedford College rather naively set out to learn what had been done in the preceding decade on groups of cohomological dimension one. This is a particularly attractive subject, that has witnessed substantial success, essentially beginning in 1968 with results of Serre, Stallings and Swan, later receiving impetus from the introduction of the concept of the fundamental group of a connected graph of groups by Bass and Serre, and recently culminating in Dunwoody's contribution which completed the characterization. Without going into definitions, one can state the result simply enough: For any nonzero ring R (associative, with 1) and group G, the augmentation ideal of the group ring R[G] is right R[G]-projective if and only if G is the fundamental group of a graph of finite groups having order invertible in R.

These notes, a (completely) revised version of those prepared for the Study Group, collect together material from several sources to present a self-contained proof of this fact, assuming at the outset only the most elementary knowledge - free groups, projective modules, etc. By making the rôle of derivations even more central to the subject than ever before, we were able to simplify some of the existing proofs, and in the process obtain a more general "relativized" version of Dunwoody's result, cf IV.2.10. An amusing outcome of this approach is that we here have a proof of one of the major results in the theory of cohomology of groups that nowhere mentions cohomology - which should make this account palatable to hard-line ring theorists. (Group theorists

PREFACE

will notice we have not touched upon the fascinating subject of ends of groups, usually one of the cornerstones of this topic, cf Cohen [72]; happily, an up-to-date outline of the subject of ends is available in the recently published lecture notes of Scott-Wall [79].)

There are four chapters. Chapter I covers, in the first six sections, the basics of the Bass-Serre theory of groups acting on trees (using derivations to prove the key theorem, I.5.3), and then in I.§8, I.§9 gives an abstract treatment of Dunwoody's results on groups acting on partially ordered sets with involution. Chapter II gives the standard classical applications of the Bass-Serre theory, including a proof of Higgins' generalization of the Grushko-Neumann theorem (based on a proof by I.M.Chiswell). Chapter III presents the Dunwoody-Stallings decomposition of a group arising from a derivation to a projective module, and gives Dunwoody's accessibility criteria. Finally, in Chapter IV, the groups of cohomological dimension one are introduced and characterized; the final section describes the basic consequences for finite extensions of free groups.

A reader interested mainly in the projectivity results of IV.§2 can pursue the following course:Chapter I:§§1-6,§8,§9; Chapter II:3.1,3.3,3.5; Chapter III:1.1,1.2,§2,§3,4.1-4.8,4.11, Chapter IV:§1,§2.

Since the subject is quite young, and the notation to some extent still tentative, we have felt at liberty to introduce new terminology and notation wherever it suited our needs, or satisfied our category-theoretic prejudices. At these points, we have made an effort to indicate the notations used by other authors.

PREFACE

Through ignorance, we have been unable to give much in the way of historical remarks, and those we have given may be inaccurate, since, as both Cohen and Scott have remarked, it is difficult to attribute, with any precision, results which existed implicitly in the literature before being made explicit.

The computer microfilm drawings, pp 13, 25, were produced by the CDC 7600 at the University of London Computer Centre, using their copyrighted software package DIMFILM. I thank Chris Cookson and Phil Taylor for their helpful technical advice in using this package.

I thank all the participants of the Study Group for their kind indulgence in this project, and especially Yuri Bahturin and Bill Stephenson for relieving me (and the audience) by giving many of the seminars.

I gratefully acknowledge much useful background material (and encouragement) from the experts at Queen Mary College, I.M. Chiswell and D.E. Cohen for Chapters I-II and III-IV respectively.

Bedford College
London
January 1980

Warren Dicks

NOTATION AND CONVENTIONS

The following notation will be used:

ϕ	for the empty set;
$\mathbb{Z}$	for the ring of integers;
$\mathbb{Q}$	for the field of rational numbers;
$\mathbb{C}$	for the field of complex numbers;
$A - B$	for the set of elements in A not in B;
$|A|$	for the cardinal of A;
B^A	for the set of all functions from A to B, the elements thought of as A-tuples with entries chosen from B;
$A \times B$, $\prod_{\alpha \in A} B_\alpha$	for the Cartesian product;
$A \vee B$, $\vee_{\alpha \in A} B_\alpha$	for the disjoint union of sets;
$A \oplus B$, $\oplus_{\alpha \in A} B_\alpha$	for the direct sum of modules.

Functions are usually, but not always, written on the right of their arguments.

All theorems, propositions, lemmas, corollaries, remarks and conventions are numbered consecutively in each section, thus 4.3 CONVENTION follows 4.2 DEFINITION in section I.4 (and outside Chapter I they are referred to as I.4.3 and I.4.2). The end of each subsection is indicated by $\square$.

References to the bibliography are by author's name and the last two digits of the year of publication, thus Serre [77], with primes to distinguish publications by the same author in the same year.

CONTENTS

CHAPTER I: GROUPS ACTING ON GRAPHS 1

- I.1 Graphs 1
- I.2 Graph morphisms and coverings 4
- I.3 Group actions 7
- I.4 Graphs of groups 10
- I.5 A tree 15
- I.6 The structure theorem 21
- I.7 An example: $SL_2(\mathbb{Z})$ 24
- I.8 Fixed points 27
- I.9 Trees and partial orders 31

CHAPTER II: FUNDAMENTAL GROUPS 35

- II.1 The trivial case - free groups 35
- II.2 Basic results 38
- II.3 The faithful case 43
- II.4 Coproducts 49

CHAPTER III: DECOMPOSITIONS 55

- III.1 Decomposing a group 55
- III.2 Cuts 62
- III.3 Decomposition theorems 68
- III.4 The relationship with derivations .. 74
- III.5 Accessibility 82

CHAPTER IV: COHOMOLOGICAL DIMENSION ONE 101

- IV.1 Projective augmentation modules 102
- IV.2 Pairs of groups 107
- IV.3 Finite extensions of free groups 118

BIBLIOGRAPHY AND AUTHOR INDEX 121

SUBJECT INDEX 125

SYMBOL INDEX 127

CHAPTER I

GROUPS ACTING ON GRAPHS

1. GRAPHS

By a _graph_ X we mean a set X that is given as the disjoint union $V \vee E$ of two sets $V = V(X) \neq \phi$ and $E = E(X)$, given with two maps $\iota, \tau : E \to V$. The elements of V are called the _vertices_ of X, and the elements of E the _edges_ of X. For $e \in E$, the vertices $\iota e, \tau e$ are called the _initial_ and _terminal_ vertices of e, respectively. An edge will usually be depicted

$$\iota e \quad \underline{\quad e \quad} \quad \tau e$$

although we also allow the possibility that $\iota e = \tau e$, in which case e is called a _loop_.

Let us fix a graph, X.

For any subset S of X we write $V(S) = S \cap V(X)$, and $E(S) = S \cap E(X)$. If for each $e \in E(S)$ we have $\iota e, \tau e \in V(S)$, then we say S is a _subgraph_ of X.

For each edge e of X we define formal symbols e^1, e^{-1}, to be thought of as travelling along e the right way and the wrong way, respectively. We set $\iota e^1 = \tau e^{-1} = \iota e$, $\tau e^1 = \iota e^{-1} = \tau e$.

By a _path_ P in X is meant a finite sequence,

(1) $\qquad P = v_0, e_1^{\varepsilon_1}, v_1, \ldots, e_n^{\varepsilon_n}, v_n$

usually abbreviated $e_1^{\varepsilon_1}, e_2^{\varepsilon_2}, \ldots, e_n^{\varepsilon_n}$, where $n \geq 0$, $\varepsilon_i = \pm 1$, and $\iota e_i^{\varepsilon_i} = v_{i-1}$, $\tau e_i^{\varepsilon_i} = v_i$ for $i = 1, \ldots, n$. We shall call v_0 the

I   GROUPS ACTING ON GRAPHS

<u>initial</u> vertex of P, and v_n the <u>terminal</u> vertex of P, and say P is a <u>path from</u> v_0 <u>to</u> v_n <u>of length</u> n.

    Two elements of X are said to be <u>connected</u> if there is a path in X containing both of them. This defines an equivalence relation on X. An equivalence class of this relation is called a <u>connected component</u> of X (or simply, a <u>component</u>), and it is easily seen to be a subgraph of X. We say that X is <u>connected</u> if it has only one component.

    Let P be a path in X as in (1). We say that P is <u>reduced</u> if for each $i = 1, \ldots, n-1$, if $e_{i+1} = e_i$ then $\varepsilon_{i+1} \neq -\varepsilon_i$, that is, $\varepsilon_{i+1} = \varepsilon_i$. If P is not reduced then for some $i = 1, \ldots, n-1$, we have $e_{i+1} = e_i$ and $\varepsilon_{i+1} = -\varepsilon_i$; in this case we say that a <u>simple reduction</u> of P gives the path

$$e_1^{\varepsilon_1}, \ldots, e_{i-1}^{\varepsilon_{i-1}}, e_{i+2}^{\varepsilon_{i+2}}, \ldots, e_n^{\varepsilon_n}.$$

By successive simple reductions we can transform P to a reduced path, called the <u>reduced form</u> of P. It is in fact unique, as can be shown by induction on the length of P, noting that any two simple reductions of P either give equal paths, or each can be followed by a suitable simple reduction to give equal paths.

    A <u>circuit</u> at a vertex v of X is a reduced path from v to itself of length at least 1. A graph with no circuits is called a <u>forest</u>, and a connected forest is called a <u>tree</u>. In a tree there is, by the above, a unique reduced path between any pair of vertices; this path will be called a <u>geodesic</u> between the vertices.

    By Zorn's Lemma there is a subgraph X' of X having $V(X') = V(X)$ and maximal with the property that X' is a forest.

By maximality, no two connected components of X' can be joined by an edge of X, so two vertices connected in X must already be connected in X'. In particular, if X is connected then so is X', in which case X' is a tree, called a <u>spanning tree</u> or a <u>maximal subtree</u> of the connected graph X.

Having assembled all these definitions, let us conclude this section by giving an algebraic characterization of trees.

For any ring R and set S we write $R[S]$ for the R-bimodule freely generated by the R-centralizing set S. Thus $R[S] = \bigoplus_{s \in S} Rs$, with $r.s = s.r$ for all $r \in R$, $s \in S$. The elements of $R[S]$ will be expressed $\sum_{s \in S} r_s.s = \sum_{s \in S} s.r_s$, where $r_s \in R$, almost all zero.

1.1 PROPOSITION. <u>Let R be a nonzero ring, and X a graph.</u> <u>Write $E = E(X)$, $V = V(V)$ There is a sequence of R-bimodules</u>

(2) $\qquad 0 \to R[E] \xrightarrow{\partial} R[V] \xrightarrow{\varepsilon} R \to 0$

<u>determined by</u> $(e)\partial = \iota e - \tau e$, $(v)\varepsilon = 1$ $(e \in E, v \in V)$.

(i) <u>The sequence is exact at</u> $R[V]$ <u>if and only if</u> X <u>is connected</u>.

(ii) <u>The sequence is exact at</u> $R[E]$ <u>if and only if</u> X <u>is a forest</u>.

(iii) <u>The sequence is exact if and only if</u> X <u>is a tree</u>. <u>In this event, for any vertex</u> v_0 <u>of</u> X, <u>the map ∂ has an R-bimodule right inverse</u> $X(-,v_0) : R[V] \to R[E]$ <u>determined by, for</u> $v \in V$, $X(v,v_0) = \varepsilon_1 e_1 + \ldots + \varepsilon_n e_n$ <u>where</u> $e_1^{\varepsilon_1}, \ldots, e_n^{\varepsilon_n}$ <u>is the geodesic from</u> v <u>to</u> v_0 <u>in the tree</u> X.

I GROUPS ACTING ON GRAPHS

Proof. (i) The cokernel of ∂ is the R-bimodule presented on generators v that centralize R, $v \in V$; relations saying $\iota e = \tau e$ for all $e \in E$. Thus Coker ∂ is $R[C]$, where C is the set of components of X. Since R is nonzero, (i) follows.

(ii), (iii) If X is not a forest then X has some circuit $e_1^{\varepsilon_1}, \ldots, e_n^{\varepsilon_n}$ with no repeated edges, and then $\varepsilon_1 e_1 + \ldots + \varepsilon_n e_n$ is a nonzero element of Ker ∂ so (2) is not exact at $R[E]$. Conversely, if X is a forest then each component of X is a tree, and it suffices to consider the case where X itself is a tree. Here, for any edge e of X, $X(-, v_0)$ sends $e\partial$ (= $\iota e - \tau e$) to $X(\iota e, v_0) - X(\tau e, v_0) = X(\iota e, \tau e) = e$. That is, $X(-, v_0)$ is a right inverse of ∂ as desired, and this verifies all the claims. □

2. GRAPH MORPHISMS AND COVERINGS

Let Γ, X be graphs.

A <u>morphism of graphs</u> $\alpha: \Gamma \to X$ is the disjoint union of two maps $V(\alpha): V(\Gamma) \to V(X)$, $E(\alpha): E(\Gamma) \to E(X)$ which have the property that for each edge e of Γ, $\alpha(\iota e) = \iota(\alpha e)$, $\alpha(\tau e) = \tau(\alpha e)$. Thus

$$\alpha(v \overset{e}{\longrightarrow} w) = \alpha v \overset{\alpha e}{\longrightarrow} \alpha w .$$

We use the terms <u>isomorphism</u> and <u>automorphism</u> of graphs in the natural way.

For any vertex v of Γ, we define

(3) $\text{star}(v) = \{e \in E(\Gamma) | \iota e = v\} \vee \{e \in E(\Gamma) | \tau e = v\}.$

GRAPH MORPHISMS AND COVERINGS §2

We say that α is <u>locally surjective</u> if for each vertex v of Γ, the induced map $\text{star}(v) \to \text{star}(\alpha v)$ is surjective; and we define <u>locally injective</u> analogously. If α is both locally injective and locally surjective then it is said to be a <u>local isomorphism</u>. For example, the morphisms

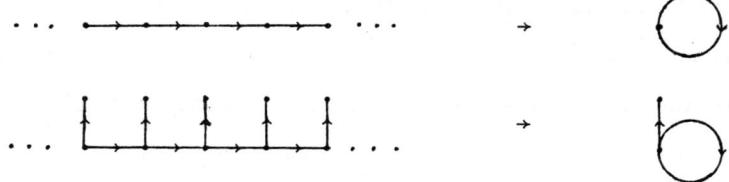

are both local isomorphisms.

2.1 PROPOSITION. <u>Let $\alpha: \Gamma \to X$ be a locally surjective graph morphism, and v be a vertex of Γ. Any subtree X' of X containing αv lifts back to a subtree Γ' of Γ containing v, that is, $\alpha: \Gamma' \to X'$ is an isomorphism.</u>

Proof. By Zorn's Lemma there exists a maximal connected subgraph Γ' of Γ containing v such that $\alpha: \Gamma' \to X'$ is injective. Notice that Γ' is then a tree. If α is not an isomorphism then there is a path P in X' starting at αv such that P does not lie entirely in $\alpha(\Gamma')$. Since X' is a tree, P traverses some edge e that does not lie in $\alpha(\Gamma')$ and has exactly one vertex in $\alpha(\Gamma')$. But as α is locally surjective, we can find a preimage of e in Γ connected to Γ', and this contradicts the maximality of Γ'. Thus α is an isomorphism. □

2.2 COROLLARY. <u>If $\alpha: \Gamma \to X$ is a locally surjective graph morphism, and X is connected then α is surjective.</u> □

I GROUPS ACTING ON GRAPHS

Notice that in the situation of 2.2 we can choose a maximal subtree T of the connected graph X and lift it back to a subtree Γ' of Γ by 2.1, and for each edge e of X not in T we can, by considering star(ιe), choose an edge f in Γ such that $\alpha f = e$ and $\iota f \in \Gamma'$. This will give us a subset S of Γ such that S is a <u>transversal</u> for α (that is, $\alpha:S \to X$ is bijective). Further, S has the property that any two vertices of S are joined by a path all of whose terms are in S, and for any edge e of S we have $\iota e \in S$; such a subset will be said to be <u>connected</u>. It is clear what we mean by a <u>maximal subtree</u> of S, so we can state the foregoing as follows.

2.3 PROPOSITION. <u>Let $\alpha:\Gamma \to X$ be a locally surjective graph morphism and X be connected. Any maximal subtree T of X lifts back to a subtree Γ' of Γ, and there exists a connected transversal S of α which has Γ' as maximal subtree.</u> □

A local isomorphism $\alpha:\Gamma \to X$ between connected graphs is called a <u>covering</u>, or a <u>covering of</u> X. The covering is said to be <u>universal</u> if Γ is a tree. By 2.2 any covering is surjective; if X is a tree we can say more.

2.4 PROPOSITION. <u>Any covering of a tree is an isomorphism.</u>

Proof. Let X be a tree and $\alpha:\Gamma \to X$ be a covering. We have seen that α is surjective so it remains to show that α is injective. Suppose that two vertices of Γ are mapped by α to a single vertex v of X. Then the reduced path between them in

GROUP ACTIONS §3

Γ is mapped to a reduced path in X from v to v, since α is locally injective. But X is a tree so the path has length zero. This shows that α is injective on vertices. Since α is locally injective, it is therefore injective on edges also. Hence α is an isomorphism. □

3. GROUP ACTIONS

Let G be a group.

We call a set X a G-<u>set</u>, or say that G <u>acts on</u> X, if there is given a group homomorphism from G to Sym_X, the group of all permutations of X. (As mappings the permutations are viewed as being written on the left of their arguments.) The image of an element g of G will usually be thought of as left multiplication by g, and denoted $x \mapsto gx$ ($x \in X$). For any $x \in X$, the <u>stabilizer of</u> x is defined to be the subgroup $G_x = \{g \in G \mid gx = x\}$, and the <u>orbit of</u> x is defined to be the G-subset $Gx = \{gx \mid g \in G\}$ of X. Notice that for $g \in G$, $x \in X$, we have $G_{gx} = gG_x g^{-1}$, so the stabilizers of two points in the same orbit are conjugate. We write h^g for $g^{-1}hg$, so $G_{gx} = G_x^{g^{-1}}$. For $x \in X$, the set of left cosets of G_x in G, $G/G_x = \{gG_x \mid g \in G\}$, is in bijective correspondence with Gx, under $gG_x \leftrightarrow gx$. This is actually an isomorphism of G-sets, if G/G_x is given its natural G-action by left multiplication. The set of orbits is denoted $G \backslash X$. There is a natural surjection $X \to G \backslash X$, $x \mapsto \bar{x} = Gx$. Choose a transversal S in X for the G-action (that is, S is a transversal for $X \to G \backslash X$). Attach to

I   GROUPS ACTING ON GRAPHS

each element of $G \backslash X$ a subgroup of G by writing, for each $x \in S$, $G(\bar{x}) = G_x$. Then as G-set, $X \simeq \bigvee_{\bar{x} \in G \backslash X} G/G(\bar{x})$.

In some situations we shall want G to act on some set X on the <u>right</u>, in which case X will be called a <u>right G-set</u>. The set of orbits is then denoted X/G, a notation which generalizes the use of G/G_x above.

Let X be a graph, and write $V = V(X)$, $E = E(X)$.

We say that G <u>acts on</u> X, or that X is a <u>G-graph</u>, if there is given a group homomorphism from G to the group of all graph automorphisms of X. Then V, E are G-sets in such a way that $g.\iota e = \iota g e$, $g.\tau e = \tau g e$ for all $e \in E$, $g \in G$. Pictorially,

$$g(\overset{v}{\bullet} \overset{e}{\longrightarrow} \overset{w}{\bullet}) = \underline{gv} \; \underline{ge} \; \underline{gw} \;.$$

3.1 EXAMPLE. Let A be a subset of G. The <u>Cayley graph</u> $\Gamma = \Gamma(G,A)$ is defined as follows: $V(\Gamma) = G$, $E(\Gamma) = G \times A$, and the incidence maps are given by $\iota(g,a) = g$, $\tau(g,a) = ga$ ($(g,a) \in E(\Gamma)$). Here G acts on Γ in a natural way by left multiplication on the vertices and on the first components of the edges. Some examples are illustrated on p. 13. We remark that $\Gamma(G,A)$ is connected if and only if A generates G. □

Let G act on X.

We write $G \backslash X$ for the graph with $V(G \backslash X) = G \backslash V$, $E(G \backslash X) = G \backslash E$, where the incidence maps are given by $\iota(Ge) = G\iota e$, $\tau(Ge) = G\tau e$, clearly well-defined.

There is then a natural surjective morphism of graphs, $X \to G \backslash X$, $x \mapsto \bar{x} = Gx$. For any vertex v of X, the induced map

GROUP ACTIONS   §3

$\text{star}(v) \to \text{star}(\bar{v})$ is surjective, for if $\bar{e} \in \text{star}(\bar{v})$, $e \in E$, then for some $g \in G$ either $g\iota e$ or $g\tau e$ equals v, so $\bar{e}$ is the image of $ge \in \text{star}(v)$. Notice that the stabilizer of v, G_v, acts on $\text{star}(v)$. Further, if e_1, e_2 are two edges with common initial vertex v, and if $\bar{e}_1 = \bar{e}_2$, then $ge_1 = e_2$ for some $g \in G$, and in fact $g \in G_v$ since $gv = g\iota e_1 = \iota e_2 = v$. This shows that the natural map $G_v \backslash \text{star}(v) \to \text{star}(\bar{v})$ is bijective.

If $G\backslash X$ is <u>connected</u>, then we may apply 2.3 to lift $G\backslash X$ back to a connected transversal in X for the G-action, call it S say. For each $e \in E(S)$ there is a unique vertex of S that lies in the same G-orbit as τe, and we may choose an element q_e of G such that $q_e^{-1} \tau e \in S$; if $\tau e \in S$ we choose $q_e = 1$. A family $(q_e \mid e \in E(S))$ chosen in this way is called a <u>connecting family</u> for S.

Each element of $G\backslash X$ can be expressed uniquely as $\bar{s}$, $s \in S$, and we can associate with it a group $G(\bar{s}) = G_s$. Then for each edge $\bar{e}$ of $G\backslash X$ ($e \in E(S)$), we have a commuting diagram

(4)
$$\begin{array}{ccc} & G(\iota\bar{e}) \to G & \\ G(\bar{e}) & & \downarrow q_e \\ & G(\tau\bar{e}) \to G & \end{array}$$

namely

(5)
$$\begin{array}{ccc} & G_{\iota e} \to G & \\ G_e & & \downarrow \\ & G_{q_e^{-1}\tau e} \to G & \end{array} \qquad \begin{array}{ccc} & g \mapsto g \\ g & & \downarrow \\ & g^{q_e} \mapsto g^{q_e} \end{array}$$

This situation will now be abstracted and studied.

I GROUPS ACTING ON GRAPHS

4. GRAPHS OF GROUPS

4.1 DEFINITION. A <u>graph of groups</u> $\mathcal{G}:Y \to \textit{Groups}$ consists of the following data: a graph Y; associated with each $y \in Y$ a group $\mathcal{G}(y)$; associated with each edge e of Y two group homomorphisms $\iota_e : \mathcal{G}(e) \to \mathcal{G}(\iota e)$, $\tau_e : \mathcal{G}(e) \to \mathcal{G}(\tau e)$. The groups $\mathcal{G}(v)$, $\mathcal{G}(e)$ ($v \in V(Y)$, $e \in E(Y)$) are called the <u>vertex groups</u> and <u>edge groups</u> of $\mathcal{G}$, respectively. We shall depict an edge of $\mathcal{G}$ as

$$\mathcal{G}(\iota e) \quad \mathcal{G}(e) \quad \mathcal{G}(\tau e) \quad .$$

If Y is connected we call $\mathcal{G}$ a <u>connected</u> graph of groups. □

By viewing a graph as a small category in a certain natural way, we see that a graph of groups is a functor from a graph to *Groups*, the category of groups and group homomorphisms. We shall often abbreviate $\mathcal{G}:Y \to \textit{Groups}$ to $(\mathcal{G},Y)$. Indeed, this is the usual notation for a graph of groups, its only disadvantage being that it does not admit an analogue for graphs of other algebraic structures such as rings.

It is clear that (4) gives us a graph of groups together with certain extra information concerning G and the q_e. We shall now see how to manufacture a situation exactly like that of (4) from any connected graph of groups.

4.2 DEFINITION. Let $(\mathcal{G},Y)$ be a connected graph of groups, and T a maximal subtree of Y. The <u>fundamental group of</u> $\mathcal{G}$ <u>with respect to</u> T, $\pi = \pi(\mathcal{G}, T)$, is the group that is universal with the following properties: associated with each $v \in V(Y)$

GRAPHS OF GROUPS §4

there is a group homomorphism $\mathcal{G}(v) \to \pi$; associated with each edge e of Y there is an element q_e of π such that the inner automorphism $q_e : \pi \to \pi$, $p \mapsto p^{q_e}$ makes the diagram

(6)
$$\begin{array}{ccc} & \mathcal{G}(\iota e) \longrightarrow & \pi \\ \mathcal{G}(e) \nearrow \searrow & & \downarrow q_e \\ & \mathcal{G}(\tau e) \longrightarrow & \pi \end{array}$$

commute, and such that $q_e = 1$ if $e \in E(T)$.

Thus π is the group presented on:

  generators the elements of $\mathcal{G}(v)$ ($v \in V(Y)$);
  generators q_e ($e \in E(Y)$);
  the relations of $\mathcal{G}(v)$ ($v \in V(Y)$);
  relations saying $(g\iota_e)^{q_e} = g\tau_e$ for all $g \in \mathcal{G}(e)$
  ($e \in E(Y)$);
  relations saying $q_e = 1$ if $e \in E(T)$. □

In the previous section we saw how a group acting on a graph with connected quotient graph gives rise to a connected graph of groups. We shall now see how a connected graph of groups gives rise to a group acting on a graph with connected quotient graph. We shall need the following.

4.3 CONVENTION. In the setting of 4.2, for each $v \in V(Y)$ we view π as a left and right $\mathcal{G}(v)$-set via the given homomorphism $\mathcal{G}(v) \to \pi$; and for each $e \in E(Y)$ as a left and right $\mathcal{G}(e)$-set via the composite $\mathcal{G}(e) \to \mathcal{G}(\iota e) \to \pi$. □

I GROUPS ACTING ON GRAPHS

4.4 DEFINITION. Let $(\mathcal{G},Y)$ be a connected graph of groups and T a maximal subtree of Y. The <u>standard graph of $\mathcal{G}$ with respect to</u> T, $\Gamma = \Gamma(\mathcal{G},T)$, is the graph with

$$V(\Gamma) = \bigvee_{v \in V(Y)} \pi/\mathcal{G}(v), \qquad E(\Gamma) = \bigvee_{e \in E(Y)} \pi/\mathcal{G}(e),$$

where the incidence maps are given by

$$\iota(p\mathcal{G}(e)) = p\mathcal{G}(\iota e), \qquad \tau(p\mathcal{G}(e)) = pq_e \mathcal{G}(\tau e)$$

($p \in \pi$, $e \in E(Y)$). This is easily seen to be well-defined, since for any $g \in \mathcal{G}(e)$, $pg\mathcal{G}(e) = pgi_e\mathcal{G}(e)$ by 4.3. □

In the next section we shall see that Γ is in fact a tree, and so Γ will be called the <u>standard tree of $\mathcal{G}$ with respect to</u> T.

Notice that π acts on Γ by left multiplication, and the quotient graph $\pi\backslash\Gamma \simeq Y$ is connected. Combined with remarks of §3 this gives us the following.

4.5 PROPOSITION (Serre [77]). <u>If a group G acts on a graph X so that G\X is connected then for any choice of connected transversal S in X for the G-action, and any choice of connecting family</u> $(q_e | e \in E(Y))$ <u>for</u> S, <u>there is a connected graph of groups</u> $G: G\backslash X \to Groups$.

<u>Conversely, if</u> $\mathcal{G}: Y \to Groups$ <u>is a connected graph of groups then for any choice of maximal subtree</u> T <u>of</u> Y <u>there is a group</u> $\pi = \pi(\mathcal{G},T)$ <u>acting on a graph</u> $\Gamma = \Gamma(\mathcal{G},T)$ <u>so that</u> $\pi\backslash\Gamma \simeq Y$. □

GRAPHS OF GROUPS §4

Group acting on graph	Graph of groups
$D_4 = \langle a,b \mid a^2 = b^2 = (ab)^4 = 1 \rangle$ 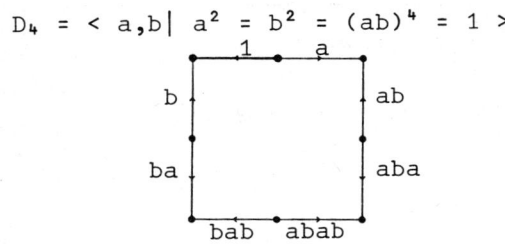	$\{1,b\} \underline{\quad 1 \quad} \{1,a\}$
$C_\infty = \langle a \mid \ \rangle$ Cayley graph: 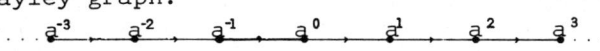	$1 \bigcirc 1$
$D_\infty = \langle a,b \mid a^2 = b^2 = 1 \rangle$ ... •ba •b •1 •a •ab •aba ...	$\{1,b\} \underline{\quad 1 \quad} \{1,a\}$
$F = \langle a,b \mid \ \rangle$ Cayley graph: 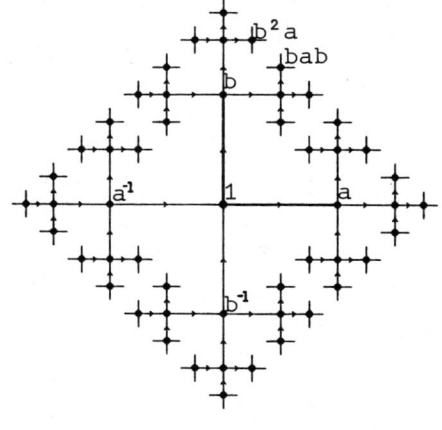	$1 \bigcirc\!\bigcirc 1$

I   GROUPS ACTING ON GRAPHS

We shall be examining $\pi(\mathcal{G},T)$ in detail in Chapter II; for the moment, let us briefly mention the two basic cases arising from one-edge graphs.

If Y has one edge and two vertices then a graph of groups can be depicted

$$A \xrightarrow{\ \ C\ \ } B$$

where A, B, C are groups and there are given homomorphisms C → A, C → B. Here the maximal subtree is the whole graph, and the fundamental group is presented on generators A∨B with relations identifying the two images of C, together with the relations of A and B. This group is denoted $A \underset{C}{\amalg} B$ called the coproduct of A and B amalgamating C. If the homomorphisms C → A, C → B are given as inclusion maps then $A \underset{C}{\amalg} B$ is called the free product of A and B amalgamating C, denoted $A \underset{C}{*} B$; we shall not be using this latter terminology and notation.

If Y has one edge and one vertex then the graph of groups can be depicted

where A, C are groups given with two homomorphisms $\alpha,\beta : C \to A$. Here the maximal subtree is the vertex, and the fundamental group is presented on generators A∨{q} with the relations of A and relations saying $q^{-1}c\alpha.q = c\beta$ for all $c \in C$. This group will be denoted HNN<α,β:C → A; q>, called the HNN construction, since it first appeared in Higman-Neumann-Neumann [49]. If α is given as an inclusion map and β is injective then

HNN$<\alpha,\beta:C \to A; q>$ is called an HNN **extension**, variously denoted A_β^* and $<A,q; q^{-1}Cq = C\beta,\beta>$ and $A_C<q;\beta>$.

5. A TREE

Throughout this section, let us fix a connected graph of groups $(\mathcal{G},Y)$ and a maximal subtree T of Y.

Write $\pi = \pi(\mathcal{G},T)$, $\Gamma = \Gamma(\mathcal{G},T)$.

Our objective now is to show that Γ is a tree. To make the notation more compact we shall write π_y in place of $\mathcal{G}(y)$, $y \in Y$, although this is dangerous since the $\mathcal{G}(y)$ are not given as subgroups of π nor is π_y the stabilizer of y in any genuine sense.

Let R be any nonzero ring. To show that Γ is a tree, it suffices by 1.1 to verify exactness of the sequence of R-bimodules

(7) $$0 \to R[E(\Gamma)] \xrightarrow{\partial} R[V(\Gamma)] \xrightarrow{\varepsilon} R \to 0$$

determined by $(e)\partial = \iota e - \tau e$, $(v)\varepsilon = 1$.

There is a natural way to make the R-bimodule $R[\pi]$ into a ring, called the **group ring**, and (7) is then a sequence of left $R[\pi]$-modules if we let π act trivially on the final (nonzero) term of (7).

For each $y \in Y$ we let $\hat{y}$ denote the element $1\pi_y$ of Γ.

Then $R[V(\Gamma)]$ is the left $R[\pi]$-module presented on

        (i) generators $\hat{v}$, $v \in V(Y)$;

        (ii) relations saying $\hat{v}$ remains fixed under the left action of π_v, where π_v acts by pullback along the homomorphism $\pi_v \to \pi$.

I   GROUPS ACTING ON GRAPHS

The left $R[\pi]$-module R is then presented with, in addition,

(iii) relations saying $\hat{\iota e} = q_e \hat{\tau e}$ for all edges e of Y.

To see this, recall that for each edge e of T, $q_e = 1$ so all the $\hat{v}$ are identified by (iii), and (ii), (iii) then say that the common image of the $\hat{v}$ is fixed by all π_v and all q_e so is fixed by π.

But $\hat{\iota e} - q_e\hat{\tau e} = \iota\hat{e} - \tau\hat{e} = (\hat{e})\partial$ so the relations (iii) hold in the cokernel of ∂ and therefore (7) is exact at $R[V(\Gamma)]$. By 1.1, Γ is therefore connected.

To complete the proof that (7) is exact and Γ is a tree, we are motivated by 1.1 to try and define for each $v,w \in V(\Gamma)$ an element $\Gamma(v,w)$ of $R[E(\Gamma)]$ that represents a geodesic; the problem is to achieve this without prior knowledge of a geodesic. What will permit us to do this is the fact that for any vertex v of Γ the map $\pi \to R[E(\Gamma)]$ $p \mapsto \Gamma(p.v,v)$ will have to have the property that $\Gamma(p_1 p_2.v, v) = p_1 \Gamma(p_2.v, v) + \Gamma(p_1.v, v)$ for all $p_1, p_2 \in \pi$. This brings us to one of the most important concepts that we shall be working with.

5.1 DEFINITION. Let G be a group and M a G-bimodule, that is, a $\mathbb{Z}[G]$-bimodule. A <u>derivation</u> $d: G \to M$ is a map such that $(ab)d = ad.b + a.bd$ for all $a, b \in G$. For example, if $m \in M$ then ad m $: G \to M$ $g \mapsto [g,m]$, where $[g,m] = gm - mg$, is a derivation, called the <u>inner derivation induced by</u> m. A derivation corresponds to a group homomorphism of the form

$$\begin{pmatrix} 1 & d \\ 0 & 1 \end{pmatrix} : G \to \begin{pmatrix} G & M \\ 0 & G \end{pmatrix}.$$

If M is given as a left G-module we can make it into a G bimodule with trivial right G-action. Here a derivation satisfies $(ab)d = (a)d + a.(b)d$; also, for any $m \in M$, ad m sends g to $(g-1)m$. Analogous statements hold for right G-modules. □

5.2 THEOREM. Let M be a π-bimodule, $(m_e \in M \mid e \in E(Y))$ <u>a family of elements</u>, and $(d_v : \pi_v \to M \mid v \in V(Y))$ <u>a family of derivations. Then the following are equivalent.</u>
(a) <u>There exists a (unique) derivation</u> $d: \pi \to M$ <u>such that for each</u> $e \in E(Y)$, $(q_e)d = m_e$, <u>and for each</u> $v \in V(Y)$ <u>the diagram</u>

$$\begin{array}{c} \pi_v \xrightarrow{d_v} M \\ \searrow \nearrow_d \\ \pi \end{array}$$

<u>commutes.</u>

(b) <u>For each</u> $e \in E(T)$, $m_e = 0$; <u>and for each</u> $e \in E(Y)$,

$$(p\iota_e)d_{\iota e} + [p\iota_e, m_e q_e^{-1}] - ((p\tau_e)d_{\tau e})q_e^{-1} = 0 \text{ for all } p \in \pi_e.$$

In particular, for any map $Y \to M$, $y \mapsto m_y$, <u>there is a (unique) derivation</u> $d: \pi \to M$ <u>such that for each</u> $e \in E(Y)$, $(q_e)d = m_e$. <u>and for each</u> $v \in V(Y)$ <u>the diagram</u>

$$\begin{array}{c} \pi_v \xrightarrow{\text{ad } m_v} M \\ \searrow \nearrow_d \\ \pi \end{array}$$

<u>commutes,</u> if and only if,

(8) <u>for each</u> $e \in E(T)$, $m_e = 0$; and
(9) <u>for each</u> $e \in E(Y)$, π_e <u>centralizes</u> $q_e m_{\tau e} q_e^{-1} - m_e q_e^{-1} - m_{\iota e}$.

Proof. It is clear that (a) is equivalent to:
(a') There is a (unique) homomorphism $\begin{pmatrix} 1 & d \\ 0 & 1 \end{pmatrix} : \pi \to \begin{pmatrix} \pi & M \\ 0 & \pi \end{pmatrix}$ such that each q_e is sent to $\begin{pmatrix} q_e & m_e \\ 0 & q_e \end{pmatrix}$, and for each vertex v of Y,

I GROUPS ACTING ON GRAPHS

$$\begin{array}{ccc} \pi_v & \longrightarrow & \pi \\ \begin{pmatrix}1 & d_v\\0 & 1\end{pmatrix}\Big\downarrow & & \Big\downarrow\begin{pmatrix}1 & d\\0 & 1\end{pmatrix} \\ \begin{pmatrix}\pi_v & M\\0 & \pi_v\end{pmatrix} & \longrightarrow & \begin{pmatrix}\pi & M\\0 & \pi\end{pmatrix} \end{array}$$

commutes.

Now by the universal property of π, this is equivalent to:

(b') For each edge e of T, $m_e = 0$, and for each edge e of Y,

$$\begin{array}{c} \pi_{\iota e} \xrightarrow{\begin{pmatrix}1 & d_{\iota e}\\0 & 1\end{pmatrix}} \begin{pmatrix}\pi_{\iota e} & M\\0 & \pi_{\iota e}\end{pmatrix} \longrightarrow \begin{pmatrix}\pi & M\\0 & \pi\end{pmatrix} \\ \pi_e \nearrow \searrow \qquad\qquad\qquad\qquad\qquad \Big\downarrow \begin{pmatrix}q_e & m_e\\0 & q_e\end{pmatrix} \\ \pi_{\tau e} \xrightarrow{\begin{pmatrix}1 & d_{\tau e}\\0 & 1\end{pmatrix}} \begin{pmatrix}\pi_{\tau e} & M\\0 & \pi_{\tau e}\end{pmatrix} \longrightarrow \begin{pmatrix}\pi & M\\0 & \pi\end{pmatrix} \end{array}$$

commutes. Performing the computation shows that this latter condition says that for each $p \in \pi_e$,

$$(p\iota_e d_{\iota e})^{q_e} + q_e^{-1} \cdot p\iota_e \cdot m_e - q_e^{-1} \cdot m_e \cdot p\tau_e = p\tau_e d_{\tau e}.$$

Since $p\tau_e \cdot q_e^{-1} = q_e^{-1} \cdot p\iota_e$, it follows that (b') is equivalent to (b), and the theorem is proved. □

We now return to proving the exactness of (7).

For any vertices v, w of Y, define

$$\Gamma(\hat{v}, \hat{w}) = \varepsilon_1 \hat{e}_1 + \ldots + \varepsilon_n \hat{e}_n,$$

where $e_1^{\varepsilon_1}, \ldots, e_n^{\varepsilon_n}$ is the geodesic from v to w in the tree T.

Fix a vertex v_0 of Y. For each vertex v of Y and $p \in \pi_v$, define

(10) $\Gamma(p\hat{v}_0, \hat{v}_0) = p\Gamma(\hat{v}_0, \hat{v}) + \Gamma(\hat{v}, \hat{v}_0) \quad \in R[E(\Gamma)].$

In other words, the map $\pi_v \to R[E(\Gamma)]$, $p \mapsto \Gamma(p\hat{v}_0, \hat{v}_0)$, is defined as the inner derivation induced by $\Gamma(\hat{v}_0, \hat{v})$.

For each edge e of Y, define

(11) $\quad \Gamma(q_e\hat{v}_0, \hat{v}_0) = q_e\Gamma(\hat{v}_0, \tau\hat{e}) - \hat{e} + \Gamma(\iota\hat{e}, \hat{v}_0) \in R[E(\Gamma)]$.

(It can be seen that this is the logical definition because $q_e\tau\hat{e} = \tau\hat{e}$.) We want to combine (10), (11) to define $\Gamma(p\hat{v}_0, \hat{v}_0)$ for all elements p of π, and by 5.2 it suffices to verify that (8) and (9) are satisfied. Here (8) requires that for each edge e of T, $\Gamma(q_e\hat{v}_0, \hat{v}_0) = 0$, which is easily verified to be the case, and (9) requires that for each edge e of Y, π_e centralizes (which in this case means stabilizes)

$$q_e\Gamma(\hat{v}_0, \tau\hat{e})q_e^{-1} - \Gamma(q_e\hat{v}_0, \hat{v}_0)q_e^{-1} - \Gamma(\hat{v}_0, \iota\hat{e})$$

which, by (11) and the triviality of the right q_e^{-1} action, equals $\hat{e}$, and this is indeed stabilized by π_e, as desired. Thus 5.2 applies, and $\Gamma(p\hat{v}_0, \hat{v}_0)$ is defined for all $p \in \pi$, and satisfies (10) and (11).

For any $v \in V(Y)$, $\Gamma(\_\hat{v}_0, \hat{v}_0) + \text{ad } \Gamma(\hat{v}, \hat{v}_0)$ vanishes on π_v by (10), so there is a well-defined R-linear map

$$\Gamma(\_, \hat{v}_0): R[V(\Gamma)] \to R[E(\Gamma)],$$

determined by

(12) $\quad \Gamma(p\hat{v}, \hat{v}_0) = p\Gamma(\hat{v}, \hat{v}_0) + \Gamma(p\hat{v}_0, \hat{v}_0), \quad p \in \pi, \quad v \in V(Y)$.

It just remains to check that $\Gamma(\_, \hat{v}_0)$ is a right inverse of the map ∂ in (7). For any edge $p\hat{e}$ of Γ ($p \in \pi$, $e \in E(Y)$) we have $(p\hat{e})\partial = p\iota\hat{e} - pq_e\tau\hat{e}$, and applying $\Gamma(\_, \hat{v}_0)$ then gives

I   GROUPS ACTING ON GRAPHS

$\Gamma(p\hat{\imath}e,\hat{v}_0) - \Gamma(pq_e\hat{\tau}e,\hat{v}_0)$

$= p\Gamma(\hat{\imath}e,\hat{v}_0) + \Gamma(p\hat{v}_0,\hat{v}_0) - pq_e\Gamma(\hat{\tau}e,\hat{v}_0) - \Gamma(pq_e\hat{v}_0,\hat{v}_0),$ by (12)

$= p\Gamma(\hat{\imath}e,\hat{v}_0) - p\Gamma(q_e\hat{v}_0,\hat{v}_0) - pq_e\Gamma(\hat{\tau}e,\hat{v}_0),$ since $\Gamma(\_\hat{v}_0,\hat{v}_0)$ is a derivation

$= p\hat{e},$ by (11).

Thus $\Gamma(\_,\hat{v}_0)$ is a right inverse of ∂, so (7) is exact, and, by 1.1, Γ is a tree.

5.3   THEOREM (Bass-Serre, Serre [77]). <u>For any connected graph of groups</u> $\mathcal{G}:Y \to$ <u>Groups and maximal subtree</u> T <u>of</u> Y, <u>the standard graph</u> $\Gamma(\mathcal{G},T)$ <u>is a tree</u>. □

We can view Γ as right π-set by defining $\gamma p = p^1\gamma$ for all $p \in \pi$, $\gamma \in \Gamma$. Hence (7) also gives an exact sequence of right $R[\pi]$-modules. From the definition of Γ we then have the following two exact sequences.

5.4   THEOREM (Chiswell [73],[76]). <u>The sequence of left</u> $R[\pi]$ <u>modules</u>

(13) $0 \to \underset{E(Y)}{\oplus} R[\pi/\pi_e] \overset{\partial}{\longrightarrow} \underset{V(Y)}{\oplus} R[\pi/\pi_v] \overset{\varepsilon}{\longrightarrow} R \to 0$

<u>determined by</u> $(p\pi_e)\partial = p\pi_{\iota e} - pq_e\pi_{\tau e}$, $(p\pi_v)\varepsilon = 1$ $(p \in \pi,\ e \in E(Y),$ $v \in V(Y)\ )$ <u>is exact</u>.

     Dually, <u>the sequence of right</u> $R[\pi]$-<u>modules</u>

(14) $\quad 0 \to \bigoplus_{E(Y)} R[\pi_e \backslash \pi] \xrightarrow{\partial} \bigoplus_{V(Y)} R[\pi_v \backslash \pi] \xrightarrow{\epsilon} R \to 0$

determined by $(\pi_e p)\partial = \pi_{1e} p - \pi_{\tau e} q_e^{-1} p$, $(\pi_v p)\epsilon = 1$
($p \in \pi$, $e \in E(Y)$, $v \in V(Y)$) is exact. □

The possibility of using the exact sequence to prove 5.3 was remarked by Chiswell; the above proof follows Dicks [77], [79] using new notation.

There are now several proofs of 5.3 available in the literature, cf. Serre [77], Chiswell [73], [77], [79]. The most geometric, and perhaps the simplest, is Chiswell [79], which is based on properties of universal coverings that our treatment will not be touching upon. From our viewpoint the above proof is the most appropriate since we shall be needing 5.2 for other purposes, so this way we virtually get 5.3 for free.

6. THE STRUCTURE THEOREM

We can now give a fairly complete picture of the relationship between groups acting on graphs and graphs of groups.

6.1 THEOREM (Bass-Serre, Serre [77]). <u>Let</u> G <u>be a group acting on a connected graph</u> X. <u>Choose a connected transversal</u> S <u>in</u> X <u>for the</u> G-<u>action, a connecting family</u> $(q_e | e \in E(Y))$ <u>for</u> S, <u>and a maximal subtree</u> T <u>of</u> S.

<u>Then there is a connected graph of groups</u> $\mathcal{G}: G\backslash X \to$ *Groups* <u>defined by</u> $\mathcal{G}(\bar{s}) = G_s$, $s \in S$, <u>where for each</u> $e \in E(S)$, <u>the maps</u> $G_e \to G_{1e}$, $G_{q_e^{-1}\tau e}$ <u>are given by</u> $g \mapsto g, g^{q_e}$ <u>respectively.</u>

I   GROUPS ACTING ON GRAPHS

From the group $\pi = \pi(G,\overline{T})$ there is a surjective homomorphism $\pi \to G$, $p \mapsto \hat{p}$, uniquely determined by the inclusion maps $G(\overline{v}) \to G$ together with $q_{\overline{e}} \mapsto q_e$ ($v \in V(S)$, $e \in E(S)$).

From the tree $\Gamma = \Gamma(G,\overline{T})$ there is a universal covering $\Gamma \to X$, $pG(\overline{s}) \mapsto \hat{p}.s$ ($p \in \pi$, $s \in S$) that respects the group actions.

Moreover, $\Gamma \to X$ is bijective if and only if $\pi \to G$ is bijective.

Proof. The construction of $G:G\backslash X \to \textit{Groups}$ is clear from 4.5 and the fact that $G\backslash X$ is connected if X is.

The existence of the homomorphism $\pi \to G$ is clear from (4) and the universal property of π. (Notice that since T is a subtree of S, the connecting family will have $q_e = 1$ for all edges e of T.)

The map $\Gamma \to X$ $pG(\overline{s}) \mapsto \hat{p}.s$ certainly respects the group actions, and is well-defined since the image of $G(\overline{s})$ in G stabilizes s. To see that it is a graph morphism, notice that for any $e \in E(S)$, $p \in \pi$, the edge

$$pG(\overline{\iota e}) \quad pG(\overline{e}) \quad pq_{\overline{e}}G(\overline{\tau e}) \qquad \text{in } \Gamma$$

is sent to

$$\hat{p}\iota e \quad \hat{p}e \quad \hat{p}q_e q_e^{-1}\tau e \qquad \text{in } X.$$

We must now check that this is a local isomorphism. Since the π-action commutes with 'star', it suffices to show that for any $v \in V(S)$ the map $\text{star}(1G(\overline{v})) \to \text{star}(v)$ is an isomorphism. Now
$\text{star}(1G(\overline{v}))$
$= \{pG(\overline{e}) \mid pG(\overline{\iota e}) = G(\overline{v})\} \vee \{pG(\overline{e}) \mid pq_{\overline{e}}G(\overline{\tau e}) = G(\overline{v})\}$

THE STRUCTURE THEOREM §6

$$= \{pG(\bar{e}) \mid \bar{\iota e} = \bar{v}, p \in G(\bar{v})\} \vee \{pG(\bar{e}) \mid \bar{\tau e} = \bar{v}, pq_{\bar{e}} \in G(\bar{v})\}$$

$$= G(\bar{v})\{1G(\bar{e}) \mid \bar{\iota e} = \bar{v}\} \vee G(\bar{v})\{q_{\bar{e}}^{-1}G(\bar{e}) \mid \bar{\tau e} = \bar{v}\} \qquad (\bar{e} \in E(G\backslash X))$$

is mapped bijectively to

$$G_v\{e \mid e \in E(S), \iota e = v\} \vee G_v\{q_e^{-1}e \mid e \in E(S), q_e^{-1}\tau e = v\}$$

which is easily seen to be star(v). Thus $\Gamma \to X$ is a local isomorphism, and since Γ is a tree by 5.3, $\Gamma \to X$ is a universal covering. In particular, it is surjective by 2.2.

Notice that as a map of sets, $\Gamma \to X$ is simply

$$\bigvee_{s \in S} \pi/G(\bar{s}) \quad \to \quad \bigvee_{s \in S} G/G_s .$$

Since S is nonempty, this is injective or surjective if and only if $\pi \to G$ has the corresponding property. Thus $\pi \to G$ is surjective and all the claims have been verified. □

In particular, if X is a tree then the universal covering $\Gamma \to X$ is an isomorphism by 2.4, and hence so is $\pi \to G$. Thus we have proved the following.

6.2 THE STRUCTURE THEOREM (Bass-Serre, Serre [77]). If X is a tree then $\pi \to G$ is an isomorphism. □

Notice that the only information about Γ needed to prove the structure theorem is the fact that Γ is connected, and we have not used the full force of 5.3.

We conclude this section by noting the converse of 6.2 is also true, and follows from 6.1 and 5.3.

I   GROUPS ACTING ON GRAPHS

7. AN EXAMPLE: $SL_2(\mathbb{Z})$

To illustrate the structure theorem, let us derive the classical description of the group

$$SL_2(\mathbb{Z}) = \{\begin{pmatrix} a & b \\ c & d \end{pmatrix} \mid a,b,c,d \in \mathbb{Z}, \ ad - bc = 1\}$$

following Serre [77].

This group acts on the upper half of the complex plane,

$$H = \{z \in \mathbb{C} \mid \text{Im } z > 0\},$$

as fractional linear (Möbius) transformations,

$$\begin{pmatrix} a & b \\ c & d \end{pmatrix} : H \to H, \qquad z \mapsto \frac{az + b}{cz + d}.$$

Let $S^1 = \{z \in \mathbb{C} \mid |z| = 1\}$. Any element $\begin{pmatrix} a & b \\ c & d \end{pmatrix}$ of $SL_2(\mathbb{Z})$ carries $S^1 \cap H$ to

(15)
$$\{z \in H \mid \ |z - \tfrac{ac - bd}{c^2 - d^2}| = |\tfrac{1}{c^2 - d^2}| \} \quad \text{if } c^2 \neq d^2$$

$$\{z \in H \mid \ \text{Re } z = ac - \tfrac{1}{2} \} \quad \text{if } c^2 = d^2 = 1.$$

View the arc $L = \{e^{i\theta} \mid \tfrac{\pi}{3} \leq \theta \leq \tfrac{\pi}{2}\}$ as an oriented edge with initial vertex i and with terminal vertex $\rho = \tfrac{1}{2} + i\tfrac{\sqrt{3}}{2}$. Let T denote the set of all translates of L under $SL_2(\mathbb{Z})$. We shall show that T is the geometric realization of a tree. From (15) we see that the only way an $SL_2(\mathbb{Z})$ translate of L can intersect L is for it to have an endpoint in common with L. It follows that T is the geometric realization of a graph. It is clear from (15) that the only point of T on the imaginary

AN EXAMPLE: $SL_2(\mathbb{Z})$ §7

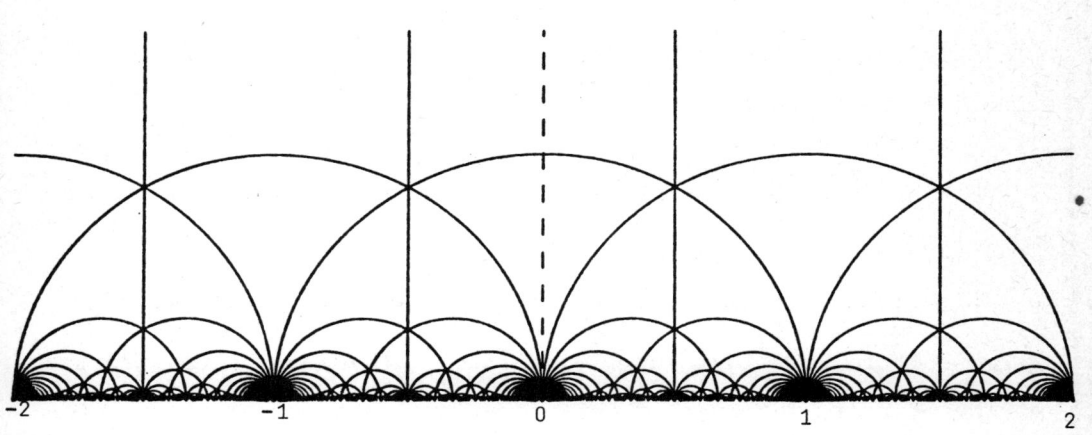

TRANSLATES OF $S^1 \cap H$ UNDER $SL_2(\mathbb{Z})$

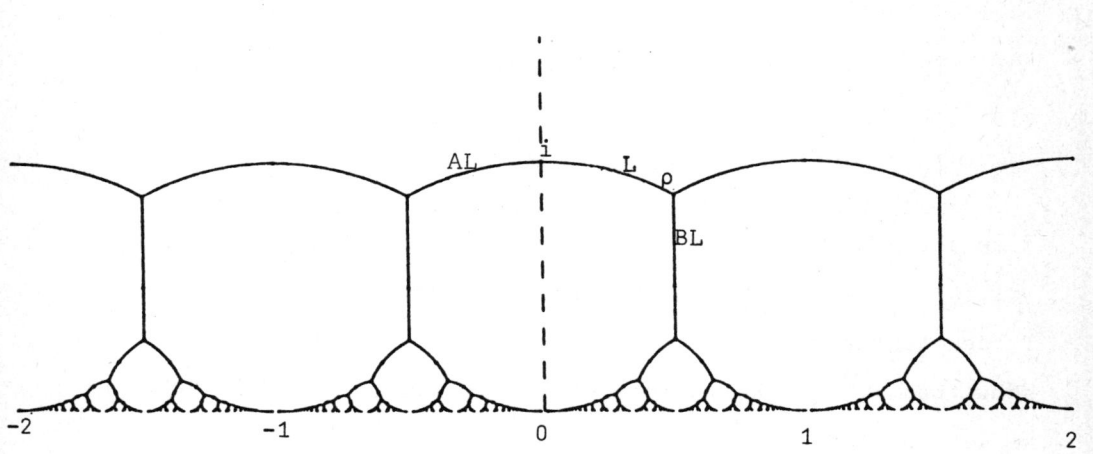

TRANSLATES OF L UNDER $SL_2(\mathbb{Z})$

I   GROUPS ACTING ON GRAPHS

axis is i, since $0 \notin H$. The only translates of L containing i are L and AL where $A = \begin{pmatrix} 0 & -1 \\ 1 & 0 \end{pmatrix}$. Hence there is no closed curve in T passing along L exactly once. But any circuit in T could be translated by a suitable element of $SL_2(\mathbb{Z})$ to a circuit that includes L. So T has no circuits. It remains to show that T is connected. By the Euclidean algorithm, any 2×2 matrix over $\mathbb{Z}$ can be transformed by the row operations $A = \begin{pmatrix} 0 & -1 \\ 1 & 0 \end{pmatrix}$, $C = \begin{pmatrix} 1 & 1 \\ 0 & 1 \end{pmatrix}$ and $C^{-1} = \begin{pmatrix} 1 & -1 \\ 0 & 1 \end{pmatrix}$ to upper triangular form. It follows that any element of $SL_2(\mathbb{Z})$ can be transformed to the identity matrix by these operations, and so $SL_2(\mathbb{Z})$ is generated by A,C and hence by A,B where $B = CA = \begin{pmatrix} 1 & -1 \\ 1 & 0 \end{pmatrix}$. Now A fixes i, and B fixes ρ, so both $L \cup AL$ and $L \cup BL$ are connected. Hence $SL_2(\mathbb{Z})L$ is connected, which completes the proof that T is a tree.

Thus $SL_2(\mathbb{Z})$ acts on the tree T, and L is a connected transversal for this action, since, by our definition of T, L contains one point from each orbit, and we have seen that it does not contain two points from any one orbit.

Computation shows that the stabilizers of L, $\iota L = i$ and $\tau L = \rho$ are generated by $-I$, A and B respectively which have orders 2, 4 and 6 respectively. So by the structure theorem, $SL_2(\mathbb{Z}) \simeq C_4 \underset{C_2}{\amalg} C_6$.

8. FIXED POINTS

Let G be a group acting on a tree, X.

We shall consider situations where G, or some element of G, fixes a vertex of X. The first of these will occur quite frequently.

8.1 THEOREM. <u>If G is finite then G fixes a vertex of</u> X.

Proof. Let v be a vertex of X and let X' be the smallest subtree of X containing the orbit Gv. Then X' is finite and is a G-subtree of X since each element of G carries X' to a subtree of X containing Gv and of the same size as X'.

Let us call a vertex of X' an (initial or terminal) <u>extremity</u> of X' if it is an (initial or terminal) vertex of exactly one edge of X'. The action of G permutes the initial extremities, and also the terminal extremities of X'. If X' has more than one vertex then deleting all the initial extremities, or deleting all the terminal extremities, together with the appropriate edges, leaves a smaller tree on which G acts. Continuing in this way, we eventually arrive at a single vertex on which G acts. □

The next results are abstracted from Dunwoody [79], many of the techniques being standard, cf. Serre [77], I.6.4, and Bass [76].

8.2 DEFINITION. Let $\geq$ be the partial order on the set

(16) $\qquad \{ e^\varepsilon \mid e \in E(X),\ \varepsilon \in \{+1,-1\} \}$

defined by saying $e_1^{\varepsilon_1} \geq e_n^{\varepsilon_n}$ for each geodesic $e_1^{\varepsilon_1},\ldots,e_n^{\varepsilon_n}$.

I GROUPS ACTING ON GRAPHS

For any element g of G and edge e of X, we say that g
shifts e if $e^\varepsilon > ge^\varepsilon$ for some $\varepsilon = \pm 1$.

8.3 LEMMA. Let g be an element of G. Then g fixes some
vertex of X if and only if g shifts no edge of X.

Proof. Suppose g fixes some vertex w of X. Let e be
any edge of X moved by g, and consider the geodesic
$e_1^{\varepsilon 1},\ldots,e_n^{\varepsilon n}$ which has $\iota e_1^{\varepsilon 1} = w$ and $e_n = e$. If i is the
least integer such that e_i is moved by g then $\iota e_i^{\varepsilon i}$ is
fixed by g and $ge_n^{-\varepsilon n},\ldots,ge_i^{-\varepsilon i},e_i^{\varepsilon i},\ldots,e_n^{\varepsilon n}$ is a reduced
path so g does not shift $e_n = e$. Conversely, if g
moves every vertex of X, choose a vertex v to minimise the
length of the geodesic $e_1^{\varepsilon 1},\ldots,e_n^{\varepsilon n}$ from v to gv. Then
$e_1^{\varepsilon 1},\ldots,e_n^{\varepsilon n},ge_1^{\varepsilon 1},\ldots,ge_n^{\varepsilon n}$ is a path from v to $g^2 v$. It is
reduced, for if $e_n^{\varepsilon n} = ge_1^{-\varepsilon 1}$ then $n > 1$ and $e_2^{\varepsilon 2},\ldots,e_{n-1}^{\varepsilon n-1}$
is the geodesic from $\tau e_1^{\varepsilon 1}$ to $\iota e_n^{\varepsilon n} = g\tau e_1^{\varepsilon 1}$, which contradicts
the minimality of n. Hence g shifts e_1. □

8.4 THEOREM. The following are equivalent.
(a) Each element of G fixes a vertex of X.
(b) No element of G shifts any edge of X.
(c) Either G fixes some vertex of X or for each vertex w of
     X there is an "infinite path" $w,e_1^{\varepsilon 1},v_1,e_2^{\varepsilon 2},\ldots$
     such that $G_{e_1} \subseteq G_{e_2} \subseteq \ldots$ is a chain of proper
     subgroups of G whose union is all of G.

Proof. (a)⇔(b) by 8.3.
(a)+(b) ⇒ (c). Consider any geodesic in X, $v_0,e_1^{\varepsilon 1},\ldots,e_n^{\varepsilon n},v_n$

Notice that either $G_{v_0} = G_{e_1}$ or $G_{e_n} = G_{v_n}$; for if not then there exist $g \in G_{v_0} - G_{e_1}$, $h \in G_{v_n} - G_{e_n}$, and then $e_1^{\varepsilon_1}, \ldots, e_n^{\varepsilon_n}, he_n^{-\varepsilon_n}, \ldots, he_1^{-\varepsilon_1}, hge_1^{\varepsilon_1}, \ldots, hge_n^{\varepsilon_n}$ is a geodesic so hg shifts e_1, contradicting (b). It follows that this geodesic has a <u>source</u> G_{v_i}, by which we mean $G_{v_0} \subseteq \ldots \subseteq G_{v_i} \supseteq \ldots \supseteq G_{v_n}$. (Although i need not be unique, here, the subgroup G_{v_i} is clearly unique.) In particular, this means that the set of vertex stabilizer subgroups forms a directed system under inclusion, so by (a), $G = \bigcup_{V(X)} G_v$.

Suppose now that G does not stabilize any vertex of X. Then for any vertex v_0 there is a vertex v such that G_{v_0} does not contain G_v so is properly contained in the source of the geodesic from v_0 to v. Hence there is an "infinite path" $w = v_0, e_1^{\varepsilon_1}, v_1, e_2^{\varepsilon_2}, \ldots$ such that $G_{v_0} \subseteq G_{v_1} \subseteq \ldots$ is an ascending chain that is not eventually constant. For any vertex v of X there is a largest n such that the geodesic from v to w passes through v_n, so v_n is the vertex of our path closest to v. For some $m > n$, $G_{v_m} \supset G_{v_n}$ so G_{v_m} must be the source for the geodesic from v to v_m, so $G_v \subseteq G_{v_m}$. Hence $\bigcup_n G_{v_n} = \bigcup_{V(X)} G_v = G$. For each n, $G_{v_n} = G_{e_n} \subseteq G_{v_{n+1}}$, and now (c) follows.

(c) $\Rightarrow$ (a) is clear. □

Let e be an edge of X. For any vertex v of X, if e is the first edge in the geodesic from ιe to v then we say e <u>points to</u> v and write $e \to v$; otherwise, e <u>points away from</u> v, denoted $v \to e$. For any vertex v of X let us write $G[e,v] = \{g \in G \mid e \to gv\}$, and let us then write

I   GROUPS ACTING ON GRAPHS

$G[e] = G[e, \iota e]$. The following properties are readily verified.

8.5 LEMMA. *Let* e *be an edge of* X, v *a vertex of* X, *and* g *an element of* G. *Then the following hold.*

(i) $(G[e,v])g = G[e, g^{-1}v]$.

(ii) $G[e,v]$ *is a right* G_v-*subset of* G.

(iii) $G[e] = \{g \in G \mid ge^1 < e^1 \text{ or } ge^1 < e^1\}$.

(iv) *For any edge* f *of* X,

$$G[e, \tau f] = \begin{cases} G[e, \iota f] \cup G_e g_0 & \text{for any } g_0 \text{ such that } g_0 f = e \\ G[e, \iota f] & \text{if } e, f \text{ are in different orbits.} \end{cases} \quad \square$$

For any subsets A, B of G, if $A - B$ is finite then we say A is almost contained in B, and write $A \stackrel{a}{\subseteq} B$. If $A \stackrel{a}{\subseteq} B$ and $B \stackrel{a}{\subseteq} A$ then A is almost equal to B, denoted $A \stackrel{a}{=} B$. If $Ag \stackrel{a}{=} A$ for every $g \in G$ then A is called almost-right-invariant. For any subgroup H of G, we say A is an almost right H-subset of G if $A \stackrel{a}{=} C$ for some right H-subset C of G.

8.6 PROPOSITION. *For any edge* e *of* X, *if* G_e *is finite then* $G[e]$ *is almost-right-invariant, and is an almost right* G_v-*subset, for every vertex* v *of* X.

Proof. For any vertex v of X, by considering the geodesic from ιe to v, we see from 8.5 (iv) that $G[e, v] \stackrel{a}{=} G[e, \iota e]$, so by 8.5 (ii), $G[e]$ is an almost right G_v-subset. Also, for any $g \in G$, $G[e, \iota e] \stackrel{a}{=} G[e, g^{-1}\iota e]$, so by 8.5 (i), $G[e] \stackrel{a}{=} G[e]g$, so $G[e]$ is almost-right-invariant. $\square$

For $g \in G$ write $\langle g \rangle$ for the cyclic subgroup generated by g.

8.7 LEMMA. *If an element* g *of* G *shifts an edge* e *of* X *then* G[e] *is not an almost right* $\langle g\rangle$-*subset*.

Proof. On replacing g by g^{-1} if necessary, we may assume $e^1 > ge^1$ so g has infinite order and G[e] contains g^n, $n > 0$, and does not contain g^n, $n < 0$, so G[e] is not an almost right $\langle g\rangle$-subset. □

8.8 THEOREM. *If the edge stabilizer subgroups are finite of bounded order then a subgroup* H *of* G *fixes a vertex of* X *if and only if every* G[e] *is an almost right* H-*subset*.

Proof. If H fixes a vertex of X then by 8.6 every G[e] is an almost right H-subset. Conversely, if H does not fix a vertex of X then 8.4 (c) fails, so some element of H shifts an edge e, so by 8.7, this G[e] is not an almost right H-subset. □

9. TREES AND PARTIAL ORDERS

In the preceding section we associated with a tree X a partially ordered set (16) possessing a natural order-reversing involution $e^\varepsilon \mapsto e^{-\varepsilon}$. Notice that if the orientation of X is altered then the resulting partially ordered set with involution is virtually the same. Thus (16) may be viewed as the edge set of an unoriented tree, or more precisely, a double cover thereof. Following Dunwoody [79] we shall give an abstract characterization of those partially ordered sets with

I GROUPS ACTING ON GRAPHS

involution which can arise in this way.

Let $(E,\leq,*)$ be a nonempty partially ordered set with involution such that for any $e,f \in E$, e is comparable to exactly one of f,f^*. We shall construct an unoriented forest with (double) edge set E.

Define an equivalence relation $\sim$ on E as follows. For $e,f \in E$ write $e \sim f$ if either $e = f$ or e <u>covers</u> f^*, that is, $e > f^*$ and no element of E lies strictly between e,f^*. (In interval notation, $[f^*,e] = \{f^*,e\}$.) To see $\sim$ is transitive, suppose $e \sim f$ and $f \sim g$, $e,f,g \in E$. If any two of e,f,g are equal then $e \sim g$ so we may suppose e,f,g distinct. Hence $e > f^*$, $f > g^*$.

Now e is comparable to exactly one of g,g^*. If e is comparable to g then f^* cannot be covered by both e,g. So e is comparable to g^*. If $e \leq g^*$ then $f^* < e \leq g^* < f$ which contradicts f being comparable to exactly one of f,f^*. So $e > g^*$, as desired.

In interval notation the fact that every element of E is comparable to f or f^* says

$$[g^*,e] = [g^*,f^*] \cup (f^*,e] \cup [g^*,f) \cup [f,e] = (f^*,e] \cup [g^*,f)$$
$$= \{e,g^*\}.$$

TREES AND PARTIAL ORDERS §9

So e covers g^*. It is now clear that $\sim$ is an equivalence relation.

Let X' be the unoriented graph with edge set $E/*$, that is, $\{\{e,e^*\} \mid e \in E\}$, and with vertex set $E/\sim$, where the edge $\{e,e^*\}$ has as vertices the classes of e, e^* in $E/\sim$. We can think of e, e^* as being the two possible orientations of $\{e,e^*\}$ where we view e as the oriented edge with terminal vertex the class of e in $E/\sim$.

For any unrefinable chain $e_1 > e_2 > \ldots > e_n$ in E (that is, e_i covers e_{i+1}, $i = 1, \ldots, n-1$),

$$\underset{\{e_1, e_1^*\}}{\bullet\!-\!-\!-\!-\!\bullet} \quad \ldots \quad \underset{\{e_n, e_n^*\}}{\bullet\!-\!-\!-\!-\!\bullet}$$

is a reduced path in X', and conversely, every reduced path in X' arises in this way. This shows that X' has no circuits so is an unoriented forest. It is clear that the necessary and sufficient condition for X' to be an unoriented tree is that for any $e, f \in E$ the (totally ordered) interval $[e,f]$ is finite.

This gives the claimed characterization.

For our purposes we need a genuine (oriented) tree and we can obtain this either by choosing an orientation of X' or by taking a barycentric subdivision of X'. It is the latter that will meet our needs.

Let X be the graph with $V(X) = E/* \vee E/\sim$, $E(X) = E$, where for each $e \in E$, ιe is the class of e in $E/*$ and τe is the class of e in $E/\sim$. Since X' is an unoriented tree it is clear that X is a tree. For any unrefinable chain

I GROUPS ACTING ON GRAPHS

$e_1 > e_2 > \ldots > e_n$ in E,

$$\underset{\leftarrow}{e_1^*} \underset{\rightarrow}{e_1} \underset{\leftarrow}{e_2^*} \underset{\rightarrow}{e_2} \ldots \underset{\leftarrow}{e_n^*} \underset{\rightarrow}{e_n}$$

is a reduced path in X.

Let the set $\{e^\varepsilon \mid e \in E, \varepsilon \in \{+1, -1\}\}$ have the partial order inherited from X, denoted $\leq$ say. Then for any e, f in E,

$$e^1 < f^{-1} \text{ if and only if } e \leq f^* \text{ in } E,$$

$$f^{-1} < e^1 \text{ if and only if } f^* < e \text{ in } E.$$

Using 8.5 (iii) we now have the following.

9.1 THEOREM (Dunwoody [79]). <u>Let $(E, \leq, *)$ be a nonempty partially ordered set with order-reversing involution such that for each e, f in E, e is comparable under $\leq$ to exactly one of f, f^*, and the (totally ordered) interval $[e, f]$ is finite. Then E is the edge set of a certain tree X.</u>

<u>If, moreover, a group G acts on E respecting the partial order and the involution, then the action of G on E extends to an action of G on X in such a way that for each e in E, $G[e] = \{g \in G \mid ge < e \text{ or } ge^* < e\}$.</u> □

CHAPTER II

FUNDAMENTAL GROUPS

In many naturally occurring situations, information about a given group can be used to construct a tree on which the group acts, and then the structure theorem provides a description of the group as the fundamental group of a certain graph of groups. This can only be useful if something is known about the fundamental group, and the purpose of this chapter is to indicate some of the salient features. Although almost all the results here are classical, the Bass-Serre approach of using the action of the fundamental group on the standard tree makes the proofs substantially more transparent. Indeed this is where one applies the full force of the standard graph being a tree. The moral is that the real information lies in the original action of our given group on the constructed tree.

1. THE TRIVIAL CASE - FREE GROUPS

Let Y be a connected graph and T be a maximal subtree of Y.

We write $Y:Y \to$ *Groups* for the <u>trivial</u> graph of groups <u>on</u> Y that assigns to each element of Y the trivial group, and to each edge of Y the identity homomorphisms. The fundamental group of Y with respect to T, $\pi(Y,T)$, has presentation

$$< q_e, \, e \in E(Y) \mid q_e = 1, \, e \in E(T) >$$

so is free of rank $|E(Y) - E(T)|$. In particular, if $E(T)$ is

II   FUNDAMENTAL GROUPS

finite then $|E(T)| = |V(Y)| - 1$ and $\pi(Y,T)$ is free of rank $|E(Y)| - |V(Y)| + 1$. Thus if Y has only one vertex, so is a bouquet of loops, then $\pi(Y,T)$ is free of rank $|E(Y)|$. Since any cardinal can occur here, we deduce that a group is free if and only if it is the fundamental group of a trivial connected graph of groups.

The trivial graph of groups on Y arises when we apply I.4.5 to the action of the trivial group on Y. Then by I.6.1 there is a universal covering $\Gamma(Y,T) \to Y$; also $\pi(Y,T)$ acts freely on $\Gamma(Y,T)$, where a group is said to act <u>freely</u> on a set if every point has trivial stabilizer. Further, $\pi(Y,T)\backslash\Gamma(Y,T) \simeq Y$.

We have thus shown that if a group is the fundamental group of a trivial connected graph of groups then it acts freely on some tree; the converse holds by I.6.2, and we have arrived at the oldest form of the Bass-Serre theorems.

1.1 THEOREM (Reidemeister [32]). <u>A group</u> G <u>is free if and only if it acts freely on some tree</u>, X <u>say</u>. <u>In this event</u>, $G \simeq \pi(G\backslash X, T)$ <u>for any maximal subtree</u> T <u>of</u> $G\backslash X$. □

Let us record one consequence of this.

1.2 THEOREM (Schreier [27]). <u>A subgroup</u> H <u>of a free group</u> G <u>is again free</u>. <u>If moreover</u> rank G <u>and</u> $(G:H)$ <u>are finite then</u>
$$\text{rank } H = (G:H)(\text{rank } G - 1) + 1.$$

Proof. Let G be free on a set E, and let Y be the bouquet of loops with edge set E. Then the maximal subtree T of Y is

THE TRIVIAL CASE - FREE GROUPS §1

just the unique vertex of Y, and $G \simeq \pi(Y,T)$ in a natural way so acts freely on the tree $X = \Gamma(Y,T)$. Clearly H acts freely on X so is free by 1.1. If moreover (G:H) is finite then as H-set, H\X is (G:H) copies of G\X, so if rank G is also finite then Y = G\X is finite and

$$\begin{aligned}
\text{rank } H &= |E(H\backslash X)| - |V(H\backslash X)| + 1 \\
&= (G:H)|E(G\backslash X)| - (G:H)|V(G\backslash X)| + 1 \\
&= (G:H)(\text{rank } G - 1) + 1. \quad \square
\end{aligned}$$

Combining 1.1 with the concepts of Chapter I we get a very useful result.

1.3 THEOREM (Bass-Serre, Serre [77]). <u>Let</u> $\mathcal{G}:Y \to$ *Groups* <u>be a graph of groups and</u> H <u>any subgroup of</u> $\pi(\mathcal{G},T)$ <u>which has trivial intersection with each conjugate of the image of each vertex group,</u> $p^{-1}\mathcal{G}(v)p$ $(p \in \pi(\mathcal{G},T), v \in V(Y))$. <u>Then</u> H <u>is free, and in fact</u> $H \simeq \pi(H\backslash\Gamma(\mathcal{G},T),T')$ <u>for any maximal subtree</u> T' <u>of</u> $H\backslash\Gamma(\mathcal{G},T)$.

Proof. The hypothesis ensures that H acts freely on the tree $\Gamma(\mathcal{G},T)$. $\square$

1.4 COROLLARY (Bass-Serre, Serre [77]). <u>If each vertex group</u> $\mathcal{G}(v)$ <u>is torsion then every torsion-free subgroup of</u> $\pi(\mathcal{G},T)$ <u>is free.</u> $\square$

1.5 REMARKS (Bass-Serre, Serre [77]). In the setting of I.6.1 with a group G acting on a connected graph X, the kernel N of $\pi(G,\overline{T}) \to G$ satisfies the hypothesis of 1.3 and further,

II FUNDAMENTAL GROUPS

$N\backslash\Gamma(G,\overline{T}) \simeq X$ in a natural way, so we have an exact sequence $1 \to \pi(X,T') \to \pi(G,\overline{T}) \to G \to 1$, for any maximal subtree T' of X.

Let us note also that for any graph of groups $\mathcal{G}:Y \to \textit{Groups}$ we have a surjection $\pi(\mathcal{G},T) \to \pi(Y,T)$ and the kernel is the normal subgroup generated by the images of the vertex groups. This says that if a group G acts on a tree X then there is a surjection $G \to \pi(G\backslash X,\overline{T})$ and the kernel is the subgroup of G generated by the stabilizers of the vertices of X. □

2. BASIC RESULTS

Let $\mathcal{G}:Y \to \textit{Groups}$ be a connected graph of groups, T be a maximal subtree of Y and v_0 a vertex of Y.

A tool that will be useful in our initial study of the fundamental group is a concept that we shall call specialization; it arises in a certain way from the category theoretic notion of a natural transformation between functors, but we shall not digress to make precise the connection.

2.1 DEFINITION. For any group G, a <u>specialization</u> from $\mathcal{G}$ to G, $t:\mathcal{G} \to G$, will consist of the following data:

a family $(t_e \mid e \in E(Y))$ of elements of G,

a family $(t_v:\mathcal{G}(v) \to G \mid v \in V(Y))$ of homomorphisms,

with the property that for each edge e of Y

(1)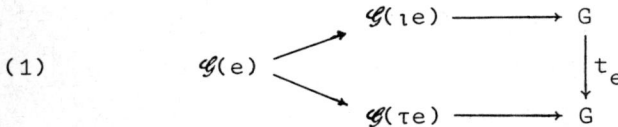

BASIC RESULTS §2

commutes, where the vertical arrow is the inner automorphism, as usual. ☐

For example, there is a specialization $t: \mathcal{G} \to \pi(\mathcal{G},T)$ with $t_e = q_e$ for every edge e of Y.

2.2 PROPOSITION (Bass-Serre, Serre [77]). <u>Let $t: \mathcal{G} \to G$ be a specialization. For any vertices v,w of Y define</u> $T(v,w) = t_{e_1}^{\varepsilon_1} \ldots t_{e_n}^{\varepsilon_n}$ <u>where</u> $e_1^{\varepsilon_1}, \ldots, e_n^{\varepsilon_n}$ <u>is the geodesic from v to w in the tree T. Then there is a unique homomorphism</u> $\pi(\mathcal{G},T) \to G$ <u>such that</u>

$$q_e \mapsto T(v_0, \iota e) t_e T(\tau e, v_0) \quad \text{<u>for each edge</u> } e \text{ <u>of</u> } Y, \text{ <u>and</u>}$$

$$\begin{array}{ccc} \mathcal{G}(v) & \longrightarrow & G \\ \downarrow & & \downarrow T(v,v_0) \\ \pi(\mathcal{G},T) & \longrightarrow & G \end{array}$$

<u>commutes for each vertex v of Y.</u>

Proof. For each edge e of T, $T(v_0, \iota e) t_e T(\tau e, v_0) = T(v_0, v_0) = 1$, and for each edge e of Y,

$$\begin{array}{ccc}
& \mathcal{G}(\iota e) \longrightarrow G \xrightarrow{T(\iota e, v_0)} G & \\
\mathcal{G}(e) \begin{array}{c}\nearrow \\ \searrow\end{array} & \downarrow t_e & \downarrow T(v_0, \iota e) t_e T(\tau e, v_0) \\
& \mathcal{G}(\tau e) \longrightarrow G \xrightarrow{T(\tau e, v_0)} G &
\end{array}$$

commutes. The result therefore follows from the universal property of $\pi(\mathcal{G},T)$. ☐

2.3 DEFINITION. Let $\mathcal{G} \to U(\mathcal{G})$ denote the universal specialization, that is, $U(\mathcal{G})$ is the group presented on generators $(t_e \mid e \in E(Y))$ together with the $\mathcal{G}(v)$, $v \in V(Y)$,

II   FUNDAMENTAL GROUPS

and relations consisting of the relations of the $\mathcal{G}(v)$ together with relations saying that for each edge e of Y, (1) commutes.

For any path $v_0, e_1^{\varepsilon_1}, \ldots, v_{n-1}, e_n^{\varepsilon_n}, v_n$ in Y, any element of $U(\mathcal{G})$ that can be expressed in the form $g_0 t_{e_1}^{\varepsilon_1} g_1 t_{e_2}^{\varepsilon_2} \ldots t_{e_n}^{\varepsilon_n} g_n$, where for each i, g_i comes from $\mathcal{G}(v_i)$, will be called a <u>path in</u> $U(\mathcal{G})$ <u>from</u> v_0 <u>to</u> v_n.

The set of paths in $U(\mathcal{G})$ from v_0 to itself, clearly a subgroup of $U(\mathcal{G})$, is called the <u>fundamental group of</u> $\mathcal{G}$ <u>at</u> v_0, denoted $\pi(\mathcal{G}, v_0)$. This group acts in a natural way on the graph $\Gamma(\mathcal{G}, v_0)$ defined as follows:

$V(\Gamma(\mathcal{G}, v_0)) = \{P\mathcal{G}(v) \mid v \in V(Y), P \text{ a path in } U(\mathcal{G}) \text{ from } v_0 \text{ to } v\}$
$E(\Gamma(\mathcal{G}, v_0)) = \{P\mathcal{G}(e) \mid e \in E(Y), P \text{ a path in } U(\mathcal{G}) \text{ from } v_0 \text{ to } \iota e\}$

with edges $P\mathcal{G}(\iota e) \longleftarrow P\mathcal{G}(e) \xrightarrow{} Pt_e \mathcal{G}(\tau e)$,

where the convention similar to I.4.3 is understood. □

For example, if $\mathcal{G} = Y$ is trivial, then the fundamental group of Y at v_0 reduces to the usual notion of the fundamental group of a connected topological space.

By 2.2 we have a homomorphism $\pi(\mathcal{G}, T) \to U(\mathcal{G})$ and its image can be seen to lie in $\pi(\mathcal{G}, v_0)$. But in the other direction there is a homomorphism $U(\mathcal{G}) \to \pi(\mathcal{G}, T)$ sending each t_e to q_e and 'fixing' each $\mathcal{G}(v)$. In particular, each $T(v_0, \iota e) t_e T(\tau e, v_0)$ is mapped to q_e so the composite $\pi(\mathcal{G}, T) \to U(\mathcal{G}) \to \pi(\mathcal{G}, T)$ is the identity, and therefore so is $\pi(\mathcal{G}, T) \to \pi(\mathcal{G}, v_0) \to \pi(\mathcal{G}, T)$. It is straightforward to check that the composite $\pi(\mathcal{G}, v_0) \to \pi(\mathcal{G}, T) \to \pi(\mathcal{G}, v_0)$ is the identity so we have an

BASIC RESULTS

isomorphism $\pi(\mathcal{G},T) \simeq \pi(\mathcal{G},v_0)$. It is now a simple matter to obtain the following.

2.4 THEOREM (Bass-Serre, Serre [77]). <u>There is a natural pair of compatible isomorphisms</u> $\pi(\mathcal{G},T) \simeq \pi(\mathcal{G},v_0)$, $\Gamma(\mathcal{G},T) \simeq \Gamma(\mathcal{G},v_0)$. <u>In particular, the isomorphism class of the fundamental group acting on the standard graph is independent of the choice of maximal subtree</u> T <u>of</u> Y. □

Let us now consider the problem of when all the homomorphisms $\mathcal{G}(y) \to \pi(\mathcal{G},T)$, $y \in Y$, are injective. A necessary condition for this to hold is that $\mathcal{G}$ be <u>faithful</u>, that is, for each edge e of Y, the homomorphisms $\iota_e, \tau_e : \mathcal{G}(e) \to \mathcal{G}(\iota e), \mathcal{G}(\tau e)$ are injective. (This terminology violates standard category theoretic usage, but here 'faithful' seems preferable to the correct 'mono' or 'monopreserving'.) To see that this condition is also sufficient we take the less usual, but simpler, approach that is based on the following elementary fact.

2.5 LEMMA. <u>Let</u> G <u>be a group acting freely on a set</u> X <u>in two ways, say</u> $\alpha, \beta : G \to \mathrm{Sym}_X$. <u>If</u> $|G\alpha \backslash X| = |G\beta \backslash X|$ <u>then there exists a</u> $t \in \mathrm{Sym}_X$ <u>such that</u>

$$\begin{array}{c} & \mathrm{Sym}_X \\ & \nearrow{\alpha} \\ G & \downarrow t \\ & \searrow{\beta} \\ & \mathrm{Sym}_X \end{array}$$

<u>commutes</u>.

Proof. We write $_\alpha X, _\beta X$ to denote the G-set determined by α, β respectively. The fact that G acts freely on $_\alpha X$ means that $_\alpha X$ is the disjoint union of $|G \backslash _\alpha X|$ G-sets (orbits) each of which is isomorphic to the (left) G-set G. Thus $_\alpha X$ and $_\beta X$ are isomorphic left G-sets and we may choose an isomorphism

II   FUNDAMENTAL GROUPS

$t: {}_\beta X \to {}_\alpha X$.   Then t is an element of Sym_X such that $t(g\beta . x) = g\alpha . tx$ for all $g \in G$, $x \in X$.   That is, $t^{-1} . g\alpha . t = g\beta$ for all $g \in G$, as desired. □

2.6 THEOREM. <u>Let</u> $\mathscr{G}$ <u>be faithful and let</u> X <u>be a set such that for each vertex</u> v <u>of</u> Y, $|\mathscr{G}(v)|$ <u>either divides</u> |X| <u>if</u> X <u>is finite</u>, <u>or is strictly smaller than</u> |X| <u>if</u> X <u>is infinite</u>. <u>Then there exists a specialization</u> $\mathscr{G} \to \text{Sym}_X$ <u>such that in the resulting homomorphism</u> $\pi(\mathscr{G},T) \to \text{Sym}_X$, <u>the composite</u> $\mathscr{G}(v) \to \pi(\mathscr{G},T) \to \text{Sym}_X$ <u>is injective for each vertex</u> v <u>of</u> Y.

Proof. The hypothesis on X ensures that for each vertex v of Y, $|\mathscr{G}(v)|$ divides |X|, so we may choose a free action of $\mathscr{G}(v)$ on X. For each edge e of Y we then have two free actions of $\mathscr{G}(e)$ on X, and the hypothesis on X ensures that the number of orbits is the same for both actions. Thus by 2.5 there exists a $t_e \in \text{Sym}_X$ such that

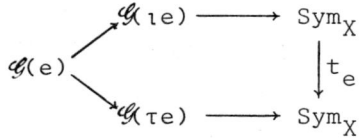

commutes. This gives the desired specialization, since the homomorphism provided by 2.2 clearly has the requisite property. □

It is clear that we can always choose a set X sufficiently large to satisfy the conditions of 2.6, so we have proved the following.

2.7 COROLLARY. *If $\mathcal{G}$ is faithful then for each vertex v of Y the homomorphism $\mathcal{G}(v) \to \pi(\mathcal{G},T)$ is injective.* □

2.8 REMARK. Together with I.§6 this last result tells us that there is a natural equivalence between, on the one hand, groups acting on trees, given with a connected transversal and a connecting family, $(G,X,S,(q_e))$, and on the other hand, faithful connected graphs of groups given with a maximal subtree, $(\mathcal{G},Y,T)$. □

3. THE FAITHFUL CASE

Let $\mathcal{G}: Y \to \textit{Groups}$ be a connected faithful graph of groups and T be a maximal subtree of Y.

Write $G = \pi(\mathcal{G},T)$ and for each $y \in Y$ write G_y for $\mathcal{G}(y)$. For each vertex v of Y the canonical map $G_v \to G$ is injective, by 2.7, and we may identify G_v with its image in G. For each edge e of Y we write G_e^{-1}, G_e^{+1} for the image of G_e under $\iota_e, \tau_e : G_e \to G_{\iota e}, G_{\tau e} \subseteq G$, respectively. Notice that the inner automorphism of G induced by q_e determines an isomorphism $G_e^{-1} \to G_e^{+1}$. If e lies in T then this map is the identity, and here we shall identify G_e with $G_e^{-1} = G_e^{+1}$.

Let $X = \Gamma(\mathcal{G},T)$. As in I.§5, there is a canonical embedding $T \to X$, $t \mapsto \hat{t}$. Let us identify T with its image in X, so the preceding conventions are consistent with stabilizer notation.

3.1 PROPOSITION. *For any edge e of Y, $G_{\iota e} \cap G_{\tau e}^{q_e^{-1}} = G_e^{-1}$, and for any vertices v, w of Y, $G_v \cap G_w$ equals the intersection of the edge groups G_e corresponding to the edges e that lie in the geodesic from v to w in T.*

II   FUNDAMENTAL GROUPS

Proof. In the action of G on the tree X, it is clear that an element of G fixes two vertices $\dot{v},w$ of $X' \supseteq T$ if and only if it fixes the geodesic between them. □

We now consider a normal form for the elements of G. Since the G_v and q_e together generate G, any element g of G can be expressed as a product

(2) $g = g_0 q_{e_1}^{\varepsilon_1} g_1 \cdots q_{e_n}^{\varepsilon_n} g_n, \quad g_i \in G_{v_i},$

for some vertices $v_0, \ldots, v_n$ and edges $e_1, \ldots, e_n$. By using geodesics in T and the identity elements from the vertex groups we may choose this expression so that $v_0, e_1^{\varepsilon_1}, \ldots, e_n^{\varepsilon_n}, v_n$ is a path P in Y. We then say that g has been expressed in path format of length n, with underlying path P.

If for some i, $e_{i+1}^{\varepsilon_{i+1}} = e_i^{-\varepsilon_i}$ in Y and $g_i \in G_{e_i}^{\varepsilon_i}$ then we can reduce the expression (2) to a path format of length n-2,

$$g_0 q_{e_1}^{\varepsilon_1} g_1 \cdots q_{e_{i-1}}^{\varepsilon_{i-1}} g' q_{e_{i+2}}^{\varepsilon_{i+2}} \cdots q_{e_n}^{\varepsilon_n} g_n$$

where $g' = g_{i-1} \cdot q_{e_i}^{\varepsilon_i} g_i q_{e_i}^{-\varepsilon_i} \cdot g_{i+1} \in G_{v_{i-1}} = G_{\iota e_i^{\varepsilon_i}} = G_{v_{i+1}}$.

On the other hand, if this does not happen for any i then the path format (2) is said to be reduced. It is clear that any element of G can be expressed in reduced path format. (Of course, the underlying path does not have to be reduced.)

We can now extend to fundamental groups a result proved by Britton [63] for the HNN construction and known as Britton's Lemma.

THE FAITHFUL CASE §3

3.2 THEOREM. <u>Let v_0 be a vertex of Y. Every element of G_{v_0} can be expressed uniquely in a reduced path format where the underlying path is from v_0 to v_0. This unique expression is the trivial path format, of length 0.</u>

Proof. Let g be an element of G_{v_0} and suppose (2) is a reduced path format with underlying path from v_0 to itself. Then in X we have a path from $v_0 = 1G_{v_0}$ to itself:

$$1G_{v_0} \xrightarrow{g_0 q_{e_1}^{\varepsilon_1} 1 G_{v_1}} \quad \xrightarrow{g_0 q_{e_1}^{\varepsilon_1} g_1 q_{e_2}^{\varepsilon_2} G_{v_2}} \cdots \xrightarrow{gG_{v_n}}$$

But X is a tree so there must be some edge followed by its inverse. On examining this edge we find there is an i such that $e_{i+1}^{\varepsilon_{i+1}} = e_i^{-\varepsilon_i}$ in Y and $g_i \in G_{e_i}^{\varepsilon_i}$. But this contradicts the path format being reduced. □

3.3 LEMMA. <u>For each vertex v of Y, let H_v be a subgroup of G_v, and suppose that for each edge e of Y, $(H_{\iota e} \cap G_e^{-1})^{q_e} = H_{\tau e} \cap G_e^{+1}$. If the H_v and q_e together generate G then $H_v = G_v$ for each vertex v of Y.</u>

Proof. For each $e \in E(Y)$ write $H_e = H_{\iota e} \cap G_e^{-1}$, and set $X' = \bigvee_{y \in Y} G/H_y$ made a graph with the (well-defined) incidence relations $\iota(gH_e) = gH_{\iota e}$, $\tau(gH_e) = gq_e H_{\tau e}$. Since the H_v, q_e generate G, X' is connected (cf p.16, or observe that X' is the image of a certain standard tree). There is an obvious locally surjective graph morphism $X' \to X$, and our hypotheses ensure that it is locally injective, so it is an isomorphism by 2.4, so $H_y = G_y$ for all $y \in Y$. □

II   FUNDAMENTAL GROUPS

3.4   COROLLARY.   If Y is finite and each edge group is finitely generated and G is finitely generated then each vertex group is finitely generated.

Proof.   Take a finite generating set for G from the generating set given by the vertex groups together with the q_e, and add in finite generating sets for each image of each edge group. For each vertex v of Y let H_v be the subgroup of G_v generated by those of our chosen generators of G that lie in G_v. Then $H_{\iota e} \supseteq G_e^{-1}$, $H_{\tau e} \supseteq G_e^{+1}$ for each edge e of Y, so by 3.3, $G_v = H_v$ is finitely generated for each vertex v of Y.   □

For the remainder of this section we shall be looking at the subgroups of G, beginning with the two extremes - finite index and finite.

3.5   THEOREM.   If H is a finite subgroup of G then H lies in some conjugate of some vertex group.

Proof.   The finite group H acts on the tree X so by I.8.1, H stabilizes a vertex of X, say g.v, g ∈ G, v ∈ V(Y). Then $H \subseteq g G_v g^{-1}$ as desired.   □

3.6   THEOREM (Karrass-Pietrowski-Solitar [73]).   If the vertex groups are all finite and of bounded order then G has a free subgroup of finite index.

Proof.   Let A be a finite set whose cardinal is the lowest common multiple of the orders of the vertex groups. By 2.6, we have a homomorphism G → Sym_A whose kernel does not meet any

THE FAITHFUL CASE §3

conjugate of any vertex group, so is free by 1.3, and clearly has finite index. □

One of the main results we shall be proving is that, conversely, every group with a free subgroup of finite index can be expressed as the fundamental group of a connected graph whose vertex groups are finite of bounded order.

Let us note the following generalization of the Schreier index formula, 1.2.

3.7 THEOREM (Serre [77]). *If Y is finite and the vertex groups are finite then for any free subgroup F of finite index in G,*

$$\frac{\text{rank } F - 1}{(G:F)} = \sum_{E(Y)} \frac{1}{|G_e|} - \sum_{V(Y)} \frac{1}{|G_v|}$$

Proof. Since F does not meet any conjugate of any vertex group, F is isomorphic to the fundamental group of $F \backslash X$ by 1.1. But the number of edges of $F \backslash X$ is

$$\sum_{E(Y)} |F \backslash G / G_e| = \sum_{E(Y)} \frac{(G:F)}{|G_e|} ,$$

and similarly for vertices. Now rank $F = |E(F \backslash X)| - |V(F \backslash X)| + 1$ gives the desired result. □

We have seen statements about free subgroups and finite subgroups of G. More generally, we can consider an arbitrary subgroup H of G. Since H acts on the tree X, the structure theorem gives a description of H as the fundamental group of a certain connected graph of groups $H \backslash X \to$ *Groups*. Here

II   FUNDAMENTAL GROUPS

the vertex groups and edge groups are intersections of H with certain conjugates of vertex groups and images of edge groups of $\mathcal{G}$. This description actually generalizes the subgroup theorems of Kurosh [37] and H.Neumann [48]. Let us mention one case that implies these results. We first recall some terminology.

3.8 DEFINITION. If Y is a tree then $\mathcal{G}:Y \to$ *Groups* is called a <u>tree of groups</u>. Here T = Y and the fundamental group is denoted simply $\pi(\mathcal{G})$. In category theoretic language, $\pi(\mathcal{G})$ is a <u>colimit</u>, and in group theoretic language is a <u>tree product</u>. If all the edge groups are trivial then $\pi(\mathcal{G})$ is the <u>coproduct</u> (or <u>free product</u>) of the vertex groups, $\coprod_{V(Y)} G_v$. Even in the case where Y is not a tree, if all the edge groups are trivial then $\pi(\mathcal{G},T) \simeq \pi(Y,T) \amalg \pi(\mathcal{G}|T)$. □

3.9 THEOREM (Bass-Serre, Serre [77]). <u>Let</u> H <u>be a subgroup of</u> G <u>that has trivial intersection with each conjugate of each</u> G_e^{-1}, $e \in E(Y)$. <u>Then</u>

$$H \simeq F \amalg \coprod_{V(Y)} (\coprod_{H\backslash G/G_v} (H \cap g^{-1}G_v g))$$

<u>where</u> F <u>is a free group</u>, <u>namely the fundamental group of the graph</u> H\X, <u>and for each vertex</u> v <u>of</u> Y, g <u>ranges over a certain complete set of double coset representatives of</u> $H\backslash G/G_v$. □

4. COPRODUCTS

We shall now prove some classical theorems on coproducts of groups from the viewpoint of fundamental groups of trees of groups with trivial edge groups.

4.1 PREPARATORY REMARKS. Let T be a tree and e,f two edges of T with $\iota e = \iota f$, thus, •—e—•—f—• . We form the quotient graph $T/(e=f)$ with one less edge and vertex than T by identifying e with f and τe with τf. It is easy to see that this quotient graph is again a tree, made up using the three connected components of $T - \{e,f\}$ together with one more edge.

More generally, if S is an equivalence relation on $E(T)$ such that $\iota e = \iota f$ for each pair $(e,f) \in S$, then by transfinite repetition of the preceding construction we can obtain the quotient graph T/S having edge set $E(T)/S$. It follows that T/S is again a tree.

We note that any automorphism of T that respects S induces an automorphism of T/S.

Of course, all these statements remain valid if ι and τ are interchanged throughout. □

4.2 THEOREM (Higgins [66]). <u>Let</u> $(G_i \mid i \in I)$, $(K_i \mid i \in I)$ <u>be two families of groups indexed by a set</u> I, <u>and write</u> $G = \coprod_I G_i$, $K = \coprod_I K_i$. <u>Suppose</u> $\alpha : G \to K$ <u>is a surjective homomorphism with</u> $(G_i)\alpha = K_i$ <u>for each</u> $i \in I$. <u>Then for any subgroup</u> H <u>of</u> G <u>such that</u> $H\alpha = K$ <u>there exists an expression</u> $H = \coprod_I H_i$ <u>with</u> $(H_i)\alpha = K_i$ <u>for each</u> $i \in I$.

II FUNDAMENTAL GROUPS

Proof (simplified version of that given in Chiswell [77], cf [76']). Clearly we may assume that I is nonempty. Let T be a tree with vertex set I; for example, a star where one specified element of I is the initial vertex of each edge. There are then natural trees of groups $\mathcal{G}, \mathcal{H} : T \to$ *Groups* with trivial edge groups, and $G = \pi(\mathcal{G})$, $K = \pi(\mathcal{H})$.

Let $X = \Gamma(\mathcal{G})$, $Z = \Gamma(\mathcal{H})$ be the standard trees on which G, K act, respectively; in particular, they act freely on the edge sets. There is a natural surjective graph morphism $X \to Z$ induced by α, and we think of Z as a quotient graph of X. (In fact, one can view Z as $N\backslash X$, where N is the kernel of α.)

Let H act on X by pullback along the inclusion map $H \to G$, and act on Z by pullback along the composite $H \to G \to K$. Thus we may view Z as a quotient H-graph of X, more precisely, a quotient H-<u>tree</u> of X.

Among the quotient H-trees Y of X that lie above Z, $X \to Y \to Z$, consider those for which H acts freely on the edge set; for example, $Y = X$. By Zorn's Lemma there exists a terminal such Y. We then have a natural surjection

$$H\backslash Y \to H\backslash Z = K\backslash Z = T.$$

Suppose that this is an isomorphism. We can then lift the canonical copy of T in Z back to a connected transversal S in Y for the H-action. Because H acts freely on $E(Y)$ the structure theorem implies $H = \coprod_{V(S)} H_i$ with $V(S) \simeq V(T) = I$. Finally, for each $i \in I$, $(H_i)\alpha = K_i$ since

COPRODUCTS §4

H_i is the H-stabilizer of the correct vertex of Z.

Thus it suffices to derive a contradiction from the supposition that $H\backslash Y \to T$ is not an isomorphism, or equivalently, is not injective on edges. Say Hy, Hy' are two edges of $H\backslash Y$ that are mapped to the same edge of T, where $y,y' \in E(Y)$.

We seek two edges of Y having a common vertex, and having common image in Z, but lying in different H-orbits.

Replacing y by a suitable H-mutiple, we may assume y and y' are mapped to the same edge of Z, z say. Let $P = y_1^{\varepsilon_1},\ldots,y_n^{\varepsilon_n}$ be a geodesic in Y having $y_1 = y$, $y_n = y'$. Clearly $n \geq 2$; we want $n = 2$. The image of P in Z, $z_1^{\varepsilon_1},\ldots,z_n^{\varepsilon_n}$, has a repeated edge $z_1 = z = z_n$ and Z is a tree so there must be some i with $z_{i+1}^{\varepsilon_{i+1}} = z_i^{-\varepsilon_i}$. If y_i and y_{i+1} are in different H-orbits then y_i, y_{i+1} is a pair of edges of Y with the desired property. Otherwise, $hy_i = y_{i+1}$ for some element h of H. This means $hz_i = z_{i+1} = z_i$ so h stabilizes an edge of Z so stabilizes all of Z since K acts freely on E(Z). (That is, $h\alpha = 1$.) Consider a new path in Y, $P' = hy_1^{\varepsilon_1},\ldots,hy_i^{\varepsilon_i},y_{i+1}^{\varepsilon_{i+1}},\ldots,y_n^{\varepsilon_n}$. Here $hy = hy_1$, and $y' = y_n$ lie in different H-orbits and have common image $hz = z$ in Z, but as P' is <u>not</u> a geodesic so hy, y' is a pair for which n is reduced by at least 2. Continuing in this way eventually gives a pair of edges of Y, y,y', having common image in Z and having common initial or terminal vertex, but lying in different H-orbits.

Now $\{(hy,hy'),(hy',hy),(e,e) \mid h \in H, e \in E(Y)\}$ is an equivalence relation on E(Y); for it is clearly reflexive and

II   FUNDAMENTAL GROUPS

symmetric, and it is vacuously transitive because y, y' lie in different H-orbits and have trivial H-stabilizers. By 4.1 the resulting quotient graph Y' of Y is a tree, in fact a quotient H-tree of Y lying above Z. Further, it is easy to see that H acts freely on $E(Y')$, so we have obtained a contradiction to the fact that Y is terminal with these properties. □

4.3 COROLLARY (Wagner [57]). <u>Let</u> F <u>be a free group</u>, $K = \coprod_I K_i$ <u>be a coproduct, and</u> $\beta: F \to K$ <u>be a surjective homomorphism.</u> <u>Then there exists an expression</u> $F = \coprod_I F_i$ <u>such that</u> $(F_i)\beta = K_i$ <u>for each</u> $i \in I$.

Proof. Let X be a free generating set for F, and for each $x \in X$ choose an expression $x\beta = k_{x,1} \cdots k_{x,n_x}$ with each $k_{x,m}$ lying in some $K_i - \{1\}$. Let $Y = \{y_{x,m}\}$ be a set indexed by the pairs (x,m), $1 \le m \le n_x$, $x \in X$. Let G be the free group on Y. Then there is a monomorphism $F \to G$ sending each $x \in X$ to $y_{x,1} \cdots y_{x,n_x}$, and we view F as a subgroup of G. For each $i \in I$ let G_i be the subgroup of G freely generated by those $y_{x,m}$ for which $k_{x,m}$ lies in K_i. Then there is a homomorphism $\alpha: G \to K$ sending each $y_{x,m}$ to $k_{x,m}$ so $(G_i)\alpha \subseteq K_i$ for each $i \in I$, and $F\alpha = K$. Thus $(G_i)\alpha = K_i$ by 3.3. Since the restriction of α to F coincides with β, the result now follows from 4.2. □

Recall that the <u>rank</u> of a group is the smallest number (cardinal) of elements required to generate it.

COPRODUCTS

4.4 THEOREM (Grushko [40], B.H.Neumann [43]). <u>For any groups</u> A,B, $\text{rank}(A \amalg B) = \text{rank } A + \text{rank } B$.

Proof. Clearly $\text{rank}(A \amalg B) \leq \text{rank } A + \text{rank } B$. Let F be the free group having the same rank as $A \amalg B$, so there is a surjective homomorphism $\beta: F \to A \amalg B$. By 4.3, $F = F_A \amalg F_B$ with $(F_A)\beta = A$, $(F_B)\beta = B$. Hence

$$\text{rank}(A \amalg B) = \text{rank } F = \text{rank } F_A + \text{rank } F_B$$
$$\geq \text{rank } A + \text{rank } B. \quad \square$$

CHAPTER III

DECOMPOSITIONS

This chapter explores the curious relationship between derivations and groups acting on trees, a relationship exemplified by a theorem of Stallings [68] which, when set into the Bass Serre theory, says that any derivation on a finitely generated group gives rise in some way to a tree on which the group acts. An actual description of such a tree, given by Dunwoody [79], permits quite a detailed analysis of the group and how it decomposes to admit the derivation. The intermediate link that will connect derivations to trees is the concept of an almost invariant set.

1. DECOMPOSING A GROUP

Let G be a group.

Suppose we are given, for a connected graph Y, a family $(G_v | v \in V(Y))$ of subgroups of G and a family $(q_e | e \in E(Y))$ of elements of G, such that $\{e | q_e = 1\}$ is the edge set of a maximal subtree T of Y. For each edge e of Y we set $G_e = G_{\iota e} \cap G_{\tau e}^{q_e^{-1}}$. Then we have a connected graph of groups $\mathcal{G}: Y \to Groups$ defined by $\mathcal{G}(y) = G_y$ ($y \in Y$) with the maps $\iota_e, \tau_e : \mathcal{G}(e) \to \mathcal{G}(\iota e), \mathcal{G}(\tau e)$ given by $g \mapsto g, g^{q_e}$, respectively. There is then a natural map $\pi(\mathcal{G},T) \to G$. If this is an isomorphism, then we call the system $(G_v, q_e | v \in V(Y), e \in E(Y))$ a Y-decomposition of G. The G_v, G_e are called the **vertex groups**

III   DECOMPOSITIONS

and <u>edge groups</u> of the decomposition; they are the data of interest in any decomposition, and we shall usually speak loosely of "a decomposition $(G_v|\ v \in V\ )$" and tacitly assume that the structure of Y and the specification of the q_e are understood.

For example {G} is a decomposition of G, called the <u>trivial</u> decomposition.

For another example, II.4.2 is really a statement about decompositions.

Associated with any decomposition of G there is a well-defined tree on which G acts, by I.5.3. Conversely, suppose G acts on a tree X and that we are given an arbitrary transversal S in X for the G-action. Then we can choose a family $(g_v|\ v \in V(S)\ )$ of elements of G so that $S' = (g_v v|\ v \in V(S)\ )$ is a <u>connected</u> transversal for the G-action, and then choose a connecting family $(q_e|\ e \in E(S)\ )$ for S'. We then have a G\X-decomposition of G, $(g_v G_v g_v^{-1}, q_e|\ v \in V(S),\ e \in E(S)\ )$. Moreover, for any specified vertex v of S we may choose $g_v = 1$, that is, we can arrange for G_v to be one of the vertex groups of the resulting decomposition.

This correspondence allows us to work with trees to obtain results about decompositions.

To state the next result it is convenient to have some more notation. If a subgroup H of G acts on a set X, let $G \otimes_H X$ denote the <u>induced</u> G-set whose elements are equivalence classes of ordered pairs, $g \otimes x$, $g \in G$, $x \in X$, with defining relations $gh \otimes x = g \otimes hx$ for all $h \in H$. Here the map $X \to G \otimes_H X$, $x \mapsto 1 \otimes x$, preserves stabilizers and induces a bijection $H \backslash X \to G \backslash (G \otimes_H X)$. There are two ways of constructing $G \otimes_H X$ directly - either choose a transversal S in X for the H-action and take the G-set $\bigvee_S G/H_s$, or choose a transversal S in G for the right H-action and make S × X a left G-set by $g(s,x) = (s',hx)$ where $gs = s'h$, $g \in G$, $s,s' \in S$, $h \in H$, $x \in X$.

1.1 THEOREM. <u>Let X be a G-tree, S a transversal in X for the G-action and suppose there is given for each $v \in V(S)$ a G_v-tree X_v. If for each edge e of X, G_e is finite then there exists a G-tree $\hat{X}$ consisting of the G-forest $\bigvee_{V(S)} G \otimes_{G_v} X_v$, together with a copy of the G-set $E(X)$.</u>

Proof. Define G-sets

$$V(\hat{X}) = \bigvee_{V(S)} G \otimes_{G_v} V(X_v) \quad \text{and} \quad E(\hat{X}) = E(X)_0 \vee \bigvee_{V(S)} G \otimes_{G_v} E(X_v)$$

where $E(X)_0$ is a copy of $E(X)$ with $E(X) \leftrightarrow E(X)_0$, $e \leftrightarrow e_0$. To define the incidence maps on $g \otimes e \in G \otimes_G E(X_v)$ set $\iota(g \otimes e) = g \otimes \iota e \in G \otimes_G V(X_v)$, and similarly for τ. Thus $\hat{X}$ contains the G-forest $\bigvee_{V(S)} G \otimes_{G_v} X_v$ and it remains to link up everything using the edges of $E(X)_0$.

Every element of $E(X)_0$ can be written in the form ge_0, $g \in G$, $e \in E(S)$. Observe that there exists a $g_e \in G$ and $v_e \in V(S)$ such that $\iota e = g_e v_e$. Then $G_{g_e^{-1}e} \subseteq G_{\iota(g_e^{-1}e)}$, that is, $g_e^{-1} G_e g_e$ is a subgroup of G_{v_e}; as it is a finite subgroup it stabilizes a vertex of X_{v_e}, by I.8.1. Choose such a vertex and denote it by the formal symbol '$g_e^{-1}\iota e_0$'. Now define, for each $g \in G$,

$\iota(ge_0) = gg_e \otimes 'g_e^{-1}\iota e_0' \in G \otimes_{G_{v_e}} X_{v_e}$; this is well-defined, for if g stabilizes e_0 then $gg_e = g_e h$ where h lies in G_{v_e} and stabilizes '$g_e^{-1}\iota e_0$'. We define τ on $E(X)_0$ similarly. Thus

III   DECOMPOSITIONS

$\hat{X}$ is a well-defined graph, and from the definitions of the incidence maps it is clear that $\hat{X}$ is a G-graph. It remains to check that $\hat{X}$ is a tree. Consider the map $\hat{X} \to X$ that collapses each tree $g \otimes_{G_v} X_v$ to the vertex gv, $g \in G$, $v \in V(S)$, and sends each e_0 to e, $e \in E(X)$. The point of choosing ιe_0 to be in the tree $g \otimes_{G_v} X_v$ for which $gv = \iota e$ is that the restriction of $\hat{X} \to X$ to the smallest subgraph of $\hat{X}$ containing $E(X)_0$ is a surjective graph morphism, so we have a fibre bundle of trees over a tree, $\hat{X} \to X$. Here the inverse image of any vertex, that is, any fibre, is a tree. Any circuit in $\hat{X}$ with no repeated edges is collapsed under $\hat{X} \to X$ to a vertex so lies in a fibre, which is impossible, so $\hat{X}$ has no circuits. Similarly, if $\hat{X}$ is not connected, then some pair of components of $\hat{X}$ is such that the images in X of the components have a vertex in common. But as the fibre over that vertex is connected, the two components are connected to each other, which is impossible. So $\hat{X}$ is connected. Hence $\hat{X}$ is a tree. □

Translating this to decompositions we obtain the following.

1.2 THEOREM (Cohen [73]). *If* $(G_v | v \in V)$ *is a decomposition of G with finite edge groups, and if for each* $v \in V$ *there is given a decomposition* $(G_w | w \in W_v)$ *of* G_v, *then there exists a decomposition* $(g_w G_w g_w^{-1} | w \in \underset{V}{V} W_v)$ *of G, for some choice of elements* g_w *of* G. *If moreover the decompositions of the* G_v *each have all edge groups finite then so does the resulting decomposition of* G. □

DECOMPOSING A GROUP §1

1.3 REMARKS. Suppose X_1, X_2 are G-trees with finite edge stabilizers. Then for each vertex v_1 of X_1, G_{v_1} acts on X_2, so by 1.1, we obtain a new G-tree $\hat{X}_1$ which is a fibre bundle over X_1. Now for any subgroups H, K of G, it is not difficult to verify that there are isomorphisms of G-sets

$$(1) \qquad G/H \times G/K \simeq \bigvee_{HgK \in H\backslash G/K} G/(H \cap gKg^{-1}) \simeq G \otimes_H (G/K).$$

Hence, from the construction of $\hat{X}_1$, we see that as G-sets $V(\hat{X}_1) \simeq V(X_1) \times V(X_2)$, $E(\hat{X}_1) \simeq E(X_1) \vee (V(X_1) \times E(X_2))$. By symmetry, we have a G-tree $\hat{X}_2$ which is a fibre bundle over X_2, and as G-sets, $V(\hat{X}_2) \simeq V(X_2) \times V(X_1)$, and $E(\hat{X}_2) \simeq E(X_2) \vee (V(X_2) \times E(X_1))$. Thus $V(\hat{X}_1) \simeq V(\hat{X}_2)$ as G-sets. Curiously, the next result says that in this circumstance, $E(\hat{X}_1) \simeq E(\hat{X}_2)$; that is,

$$E(X_1) \vee (V(X_1) \times E(X_2)) \simeq E(X_2) \vee (V(X_2) \times E(X_1)). \qquad \square$$

1.4 THEOREM. <u>If X_1 and X_2 are G-trees such that $V(X_1) \simeq V(X_2)$ as G-sets, then $E(X_1) \simeq E(X_2)$ as G-sets.</u>

Proof. Let us write $V_1 = V(X_1)$, $E_1 = E(X_1)$, and similarly for X_2. Let $\alpha: V_1 \to V_2$, and consider the additive map $E(\alpha): \mathbb{Z}[E_1] \to \mathbb{Z}[E_2]$, $e_1 \mapsto X_2((\iota e_1)\alpha, (\tau e_1)\alpha)$, $e_1 \in E_1$. (Here, as usual, for $v, w \in V_2$ we are writing $X_2(v,w)$ for $\varepsilon_1 e_1 + \ldots + \varepsilon_n e_n$, where $e_1^{\varepsilon_1}, \ldots, e_n^{\varepsilon_n}$ is the geodesic in X_2 from v to w.) It is easy to show that $E(\alpha)$ is $\mathbb{Z}[G]$-linear, and that $E(\alpha)$, $E(\alpha^{-1})$ are mutually inverse.

For any $e_1 \in E_1$, $G_{e_1} = G_{\iota e_1} \cap G_{\tau e_1} \subseteq G_{(\iota e_1)\alpha} \cap G_{(\tau e_1)\alpha}$, so G_{e_1} stabilizes the geodesic in X_2 from $(\iota e_1)\alpha$ to $(\tau e_1)\alpha$.

III   DECOMPOSITIONS

Let H be any subgroup of G. Write E_1^H for the set of elements of E_1 fixed by H, and write $E_1(H)$ for the set of elements of E_1 whose stabilizer is precisely H. We have shown that $E(\alpha)$ carries E_1^H to $\mathbb{Z}[E_2^H]$, and similarly, $E(\alpha^1)$ carries $\mathbb{Z}[E_2^H]$ to $\mathbb{Z}[E_1^H]$. Hence $E(\alpha): \mathbb{Z}[E_1^H] \cong \mathbb{Z}[E_2^H]$, and similarly, $E(\alpha): \mathbb{Z}[E_1^H - E_1(H)] \cong \mathbb{Z}[E_2^H - E_2(H)]$; and these are isomorphisms of $\mathbb{Z}[N]$-modules, where N is the normalizer of H in G. Hence there is induced a $\mathbb{Z}[N]$-linear isomorphism on the quotients, $\mathbb{Z}[E_1(H)] \cong \mathbb{Z}[E_2(H)]$. Therefore $\mathbb{Z} \otimes_{\mathbb{Z}[N]} \mathbb{Z}[E_1(H)] \cong \mathbb{Z} \otimes_{\mathbb{Z}[N]} \mathbb{Z}[E_2(H)]$ as abelian groups; that is, they have the same rank. But this rank is precisely the number of orbits isomorphic to G/H. Since these numbers are the same for E_1 and E_2 no matter which subgroup H is considered, we see $E_1 \cong E_2$ as G-sets. □

We also have an analogue of Schanuel's Lemma for G-trees.

1.5 THEOREM. *Let X_1 and X_2 be G-trees such that for each vertex v_1 of X_1, G_{v_1} stabilizes a vertex of X_2, and for each vertex v_2 of X_2, G_{v_2} stabilizes a vertex of X_1. Then*
$$V(X_1) \vee E(X_2) \cong V(X_2) \vee E(X_1) \text{ as } G\text{-sets.}$$

Proof. Let us write $V_1 = V(X_1)$, $E_1 = E(X_1)$, and similarly for X_2. The hypotheses are equivalent to the existence of G-set morphisms $\alpha: V_1 \to V_2$, $\beta: V_2 \to V_1$. There is then an additive map $E(\alpha, \beta): \mathbb{Z}[V_1 \vee E_2] \to \mathbb{Z}[V_2 \vee E_1]$ defined by

$$v_1 \mapsto v_1\alpha + X_1(v_1, v_1\alpha\beta), \quad v_1 \in V_1; \text{ and}$$
$$e_2 \mapsto (\iota e_2 - \tau e_2) + X_1((\tau e_2)\beta, (\iota e_2)\beta), \quad e_2 \in E_2.$$

It is straightforward to show that $E(\alpha, \beta)$ and $E(\beta, \alpha)$ are mutually inverse $\mathbb{Z}[G]$-linear maps, and the argument used in the proof of the preceding theorem shows that $V_1 \vee E_2 \cong V_2 \vee E_1$. □

It is interesting to note that it is now clear why the right hand side of the equation in II.3.7 does not depend on the choice of the tree.

Let X be a G-tree. We say X is <u>unreduced</u> if, for some edge e of X that has ιe and τe in different orbits, one of $G_{\iota e}$, $G_{\tau e}$ contains the other. In this event, say for instance $G_{\iota e} \subseteq G_{\tau e}$, then by "contracting e to τe" we obtain a new G-tree, X', with $V(X') = V(X) - G\iota e$, $E(X') = E(X) - Ge$. If $G\backslash X$ is finite, then, after a finite number of such steps, we arrive at a reduced G-tree.

Notice that for any reduced G-tree X with finite edge stabilizers, for any $v, w \in V(X)$, if $G_v \subseteq G_w$ then, in the geodesic $v_0, e_1^{\varepsilon_1}, \ldots, e_n^{\varepsilon_n}, v_n$ in X from v to w, all the v_i lie in the same orbit, and have the same stabilizer; for, $G_{v_0} = G_{v_0} \cap G_{v_n} \subseteq G_{e_1} \subseteq G_{v_1}$, and X is reduced, so v_0, v_1 lie in the same orbit, and their stabilizers, being finite and conjugate, must be equal; and so on for $v_2, \ldots, v_n$. In particular, for any G-subset V of $V(X)$, any G-set morphism $\alpha: V \to V(X)$ is injective and has image V, since $G_v \subseteq G_{v\alpha}$ for $v \in V$.

1.6 REMARKS. Let X_1, X_2 be two G-trees with finite edge stabilizers. Let V_1 be the set of vertices v_1 of X_1 such that G_{v_1} stabilizes a vertex of X_2; and define V_2 analogously. Thus there exist morphisms of G-sets $\alpha: V_1 \to V(X_2)$, $\beta: V_2 \to V(X_1)$.

If X_2 is reduced and $V_1 = V(X_1)$ then $\beta\alpha: V_2 \to V(X_2)$ is injective with image V_2, so β is injective.

Similarly, if $V_1 = V(X_1)$, $V_2 = V(X_2)$, and X_1, X_2 are reduced then $\alpha\beta$, $\beta\alpha$ are automorphisms, so $V(X_1) \simeq V(X_2)$ as

III   DECOMPOSITIONS

G-sets, and by 1.4, $E(X_1) \simeq E(X_2)$. □

2.   CUTS

Let G be a group and S a subset of G. Write Γ for the Cayley graph $\Gamma(G,S)$.

2.1 DEFINITION. For any graph X and subgraph A of X the coboundary δA of A in X is defined to be the set of edges of X that have to be removed to disconnect A from X, that is, $\delta A = \{e \in E(X) | \, e$ has one vertex in A and one not in $A\}$. We denote by $V\delta A$ the set of all those vertices of X that belong to some edge lying in δA. Clearly δA is empty if and only if each component of A lies in a different component of X. If X is connected then δA is the disjoint union of the coboundaries of the components of A, so, in particular, A is disconnected if and only if A has a (nonempty) subgraph B such that $\delta B \subset \delta A$, and then B is a union of components of A.

A cut of X is a subgraph A such that δA is finite. Notice that if X is connected then a cut has only finitely many components, each of which is again a cut. □

For any nonempty subset A of G we denote by A_S the largest subgraph of Γ which has vertex set A, that is, an edge of Γ lies in A_S if and only if both vertices lie in A. We shall be interested in the S-coboundary of A, $\delta_S A = \delta(A_S)$; as there is no risk of confusion we shall write δA in place of $\delta_S A$. We make the convention that the empty subset of G has empty S-coboundary. Notice that for any subset A of G, the complement A^* of A in G has the same S-coboundary as A. If δA is finite then A is called an S-<u>cut</u> of G; A is called a <u>proper</u> S-cut if both A and A^* are infinite, and A is called <u>connected</u> if it is empty or if A_S is connected. To describe S-cuts we recall the following terminology from I.§8.

2.2 DEFINITION. For any subsets A,B of G, if $A - B$ is finite then we say A is <u>almost contained</u> in B, and write $A \stackrel{a}{\subseteq} B$. If $A \stackrel{a}{\subseteq} B$ and $B \stackrel{a}{\subseteq} A$ then A is <u>almost equal</u> to B, denoted $A \stackrel{a}{=} B$. A subset A of G is called <u>almost-right-invariant</u> if $Ag \stackrel{a}{=} A$ for every $g \in G$. □

2.3 PROPOSITION. <u>A subset A of G is an S-cut if and only if $As \stackrel{a}{=} A$ for every $s \in S$ and $As = A$ for almost every $s \in S$. In this event, if S generates G then A is almost-right-invariant.</u>

Proof. This is clear from the fact that A is an S-cut if and only if there are only finitely many $(a,s) \in A \times S$ with either $as \notin A$ or $as^{-1} \notin A$. □

The remainder of this section is devoted to results about S-cuts needed in the next section.

2.4 LEMMA (Cohen [70]). <u>If S generates G then for any S-cuts A,B of G, $A \stackrel{a}{=} \{a \in A | aB \subseteq A$ or $aB^* \subseteq A\}$.</u>

Proof. Clearly we may assume that A is nonempty, and here

III   DECOMPOSITIONS

(2)
$$\begin{aligned}
\{g \in G|\ gB \subseteq A\} &= \{g \in G|\ \delta(A^* \cap gB) = \phi\} \\
&\supseteq \{g \in G|\ A^* \cap V\delta gB = \phi = V\delta A^* \cap gB\} \\
&= \{g \in G|\ V\delta gB \subseteq A \text{ and } V\delta A \subseteq gB^*\}.
\end{aligned}$$

Since A is an S-cut and Γ is connected we may choose a finite connected subgraph $\overline{\delta A}$ of Γ containing δA.

Then $A \stackrel{a}{=} \{a \in A|\ aV\delta B \subseteq A\}$ since A is almost-right-invariant and $V\delta B$ is finite

$\stackrel{a}{=} \{a \in A|\ aV\delta B \subseteq A \text{ and } \overline{\delta A} \cap a\delta B = \phi\}$ since $\overline{\delta A}$ and δB are finite

$\subseteq \{a \in A|\ V\delta aB \subseteq A \text{ and } \overline{\delta A} \text{ lies in } aB_S \text{ or } aB_S^*\}$ since $\overline{\delta A}$ is connected

$\subseteq \{a \in A|\ V\delta aB \subseteq A \text{ and } V\delta A \text{ lies in } aB \text{ or } aB^*\}$

$\subseteq \{a \in A|\ aB \subseteq A \text{ or } aB^* \subseteq A\}$ by (2). □

2.5 DEFINITION. On the power set of G, $P(G)$ (that is, the set of all subsets of G) let $\preccurlyeq$ be the relation defined by setting $A \preccurlyeq B$ $(A, B \in P(G))$ if there is a permutation of the elements of G that moves only finitely many elements of G and carries A into B; clearly $\preccurlyeq$ is reflexive and transitive, that is, $\preccurlyeq$ is a preorder. We shall be more interested in the formulation $A \preccurlyeq B$ if and only if $|A - B|$ is finite and is at most $|B - A|$. In particular, if $|A - B|$ is finite then A and B are comparable under $\preccurlyeq$.

Since it is a preoder, $\preccurlyeq$ induces an equivalence relation on $P(G)$, and for each subset A of G we have an equivalence class

$$[A] = \{B \subseteq G\ |\ A \preccurlyeq B \preccurlyeq A\}.$$

Let $\bar{P}(G)$ denote the quotient set consisting of all such equivalence classes. The partial order induced on $\bar{P}(G)$ by $\preccurlyeq$ will be denoted $\leq$. Since the left action of G on $P(G)$ respects $\preccurlyeq$ there is induced a left G-action on $\bar{P}(G)$ (and similarly for the right action, but this does not concern us). The involution $*: P(G) \to P(G)$, $A \mapsto A^*$, reverses $\preccurlyeq$ so induces an order-reversing involution, $*$, on $\bar{P}(G)$. □

2.6 LEMMA. <u>Suppose</u> S <u>generates</u> G <u>and let</u> A, B <u>be proper</u> S-<u>cuts of</u> G. <u>Then</u>

   (i) $\{g \in G | \ g[B] < [A] \ \underline{or} \ g[B^*] < [A]\} \stackrel{a}{=} A$.

  (ii) $\{g \in G | \ g[B] = [A]\}$ <u>is finite</u>.

Proof. Let $C = \{g \in G | \ gB \stackrel{a}{=} A\}$. Then for any $b \in B$, $Cb \subseteq A$ so $C \subseteq Ab^{-1} \stackrel{a}{=} A$ by 2.3, and thus $C \stackrel{a}{\subseteq} A$. But we also have $C = \{g \in G | \ gB^* \stackrel{a}{=} A^*\}$, so $C \stackrel{a}{\subseteq} A^*$ and therefore C must be finite. Since this also holds with B^* in place of B, we deduce that $\{g \in G | \ gB \stackrel{a}{=} A \text{ or } gB^* \stackrel{a}{=} A\}$ is finite. So by 2.4
$$A \stackrel{a}{=} \{a \in A | \ aB \subset A \text{ or } aB^* \subset A\},$$ so a fortiori
$$A \stackrel{a}{=} \{a \in A | \ a[B] < [A] \text{ or } a[B^*] < [A]\}.$$ The corresponding result with A^* in place of A gives
$$A^* \stackrel{a}{=} \{g \in A^* | \ g[B] > [A] \text{ or } g[B^*] > [A]\}.$$ Since $g[B], g[B^*]$ are not comparable under $\leq$ we deduce
$$A \stackrel{a}{=} \{g \in G | \ g[B] < [A] \text{ or } g[B^*] < [A]\},$$ and hence (i), (ii). □

2.7 DEFINITION. Let $P_S(G)$ denote the set of proper connected S-cuts of G, and define a partial order $\gtrsim_S$ on $P_S(G)$ by setting $A \gtrsim_S B$ $(A, B \in P_S(G))$ if either $|\delta A| > |\delta B|$ or $A \supset B$

III   DECOMPOSITIONS

and $|\delta A| = |\delta B|$. A subset Φ of $P_S(G)$ will be called <u>full</u> if for every $g \in G$ and $A \in \Phi$ at least one of the four sets

(3) $A \cap gA$, $A^* \cap gA$, $A \cap gA^*$, $A^* \cap gA^*$

has all its infinite components of the form xB, $x \in G$, $B \in \Phi$, $A \gtrsim_S B$. □

The following generalizes and simplifies results of Bergman [68] and Cohen [70].

2.8 THEOREM (Dunwoody [79]). <u>Suppose</u> S <u>generates</u> G. <u>Then any finite family of proper connected</u> S-<u>cuts of</u> G <u>is contained in a finite full family</u>.

Proof. For the purposes of this argument let us define a <u>thin</u> S-cut of G to be a proper connected S-cut that contains 1. The statement of the theorem concerns orbits in $P_S(G)$ under the left G-action, and as each orbit contains thin S-cuts it suffices to consider thin S-cuts, and in fact it suffices to show that any thin S-cut belongs to a finite family of thin S-cuts.

Let us now associate with each thin S-cut A of G a finite family $\mathcal{S}(A)$ of thin S-cuts $\gtrsim_S A$ such that for each $g \in G$ at least one of the four sets (3) has all its infinite components of the form xB, $x \in B$, $B \in \mathcal{S}(A)$. We construct such an $\mathcal{S}(A)$ as follows.

We are concerned only with those $g \in G$ for which all four of the sets (3) are infinite, and there are only finitely many such bad elements of G since by 2.4,

$G \stackrel{a}{=} \{g \in G | \, gA \text{ or } gA^* \text{ is contained in } A \text{ or } A^*\}$. For each bad g we now choose some thin S-cuts to put in $\mathcal{S}(A)$.

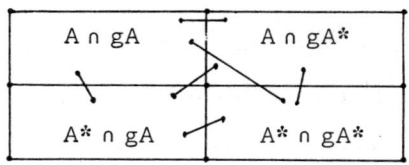

(4) $\quad |\delta(A \cap gA)| + |\delta(A^* \cap gA)| + |\delta(A \cap gA^*)| + |\delta(A^* \cap gA^*)|$

$\quad = 2|\delta(A \cap gA) \cup \delta(A^* \cap gA) \cup \delta(A \cap gA^*) \cup \delta(A^* \cap gA^*)|$

$\quad = 2|\delta A \cup \delta gA|$

$\quad \leq 4|\delta A|$.

Thus one of the four summands in (4) is $< |\delta A|$, or all four summands $= |\delta A|$. In the former case, one of the intersections (3), call it B, has $|\delta B| < |\delta A|$; for each infinite component C of B choose a $c \in C$ and put $c^{-1}C$ in $\mathcal{S}(A)$, noting that $|\delta(c^{-1}C)| \leq |\delta B| < |\delta A|$ so $c^{-1}C \underset{\mathcal{S}}{\lneq} A$. In the latter case, let B be the intersection (3) that contains 1. If B is connected then it is a thin S-cut and we put it in $\mathcal{S}(A)$. Since $B \subset A$, $B \underset{\mathcal{S}}{\lneq} A$. If B is not connected, then for each infinite component C of B choose a $c \in C$ and put the thin set $c^{-1}C$ in $\mathcal{S}(A)$, noting that $|\delta(c^{-1}C)| < |\delta B| \leq |\delta A|$, so $c^{-1}C \underset{\mathcal{S}}{\lneq} A$.

Repeating this process for each of the finitely many bad g, we obtain a family $\mathcal{S}(A)$ with the desired properties. For each thin S-cut A, call the elements of the constructed set $\mathcal{S}(A)$ the <u>chosen successors of</u> A.

For any thin S-cut A of G, consider the family Φ of thin

III   DECOMPOSITIONS

S-cuts generated by A under taking chosen successors. This is clearly a full family and it remains to show that Φ is finite. Suppose Φ is infinite. Since each thin S-cut has only finitely many chosen successors, the König Tree Lemma implies that there is in Φ an infinite sequence, $A_0, A_1, A_2, \ldots$ where for each n, A_{n+1} is a chosen successor of A_n. So in particular, $A_0 \gneq A_1 \gneq A_2 \gneq \ldots$. Hence the sequence of positive integers $|\delta A_0| \geq |\delta A_1| \geq |\delta A_2| \geq \ldots$ is eventually constant, say from the Nth term. By definition of $\gneq$ this means $A_N \supset A_{N+1} \supset \ldots$. Let $A_\omega = \bigcap_{n \geq N} A_n$. Since it contains 1 A_ω is nonempty. Also, any finite subset of δA_ω lies in δA_n for some $n \geq N$. But the $|\delta A_n|$, $n \geq N$, are all equal, so there is some $n \geq N$ such that $\delta A_\omega \subseteq \delta A_n$. Since A_n is connected this means $A_n = A_\omega$, contradicting $A_n \supset A_{n+1}$. Thus Φ is finite as desired. □

3. DECOMPOSITION THEOREMS

Let G be a group and H a subgroup of G.

In this section we examine how certain conditions give rise to decompositions of G; although these conditions appear rather technical, they occur in the study of derivations, and the results proved here will be applied in the next section to give a deep understanding of derivations.

We shall say that G is <u>finitely generated over</u> H if G has a generating set consisting of H and finitely many other elements; such a set will be said to <u>generate</u> G <u>finitely over</u> H. (It is more usual to speak of the pair (G,H) being finitely generated, but we shall not be doing so.) This concept is related to the cuts of the previous section by the following consequence of 2.3.

3.1 LEMMA. <u>Suppose</u> S <u>generates</u> G <u>finitely over</u> H. <u>Then any almost-right-invariant right</u> H-<u>subset of</u> G <u>is an</u> S-<u>cut, and its finitely many</u> S-<u>connected components are again almost-right-invariant right</u> H-<u>subsets of</u> G. □

We shall say that a given decomposition of G is <u>finitary</u> if each edge group is finite and the underlying graph is finite. A decomposition of G, $(G_v \mid v \in V)$, is said to be <u>over</u> H if there is specified a vertex $v_H \in V$ such that H is contained in G_{v_H}. For example, a decomposition over 1 is simply a decomposition with a distinguished vertex.

Recall from I.§8 that a subset A of G is called an <u>almost right</u> H-<u>subset</u> if $A \stackrel{a}{=} B$ for some right H-subset B of G. We now arrive at a preliminary form of the main result of this section - differing only in having the additional hypothesis (5).

3.2 THEOREM (Dunwoody [79]). <u>Suppose</u> G <u>is finitely generated over</u> H. <u>Let</u> Φ <u>be a finite set of almost-right-invariant right</u> H-<u>subsets of</u> G <u>such that</u>

(5) <u>for each</u> $g \in G$ <u>and each</u> $A, B \in \Phi$, <u>one of the four sets</u> $A \cap gB$, $A^* \cap gB$, $A \cap gB^*$, $A^* \cap gB^*$ <u>is finite.</u>

<u>Then there is a finitary decomposition of</u> G <u>over</u> H, $(G_v \mid v \in V)$, <u>such that for each</u> $A \in \Phi$, <u>and</u> $v \in V$, A <u>is an almost right</u> G_v-<u>subset of</u> G.

III   DECOMPOSITIONS

Proof. Let S generate G finitely over H, so the elements of Φ are S-cuts. Without loss of generality we may assume that Φ is nonempty, that the elements of Φ are __proper__ S-cuts (and right H-subsets), and that Φ is closed under the involution *.

Let $\Sigma = G\Phi$, that is, $\{gA \mid g \in G, A \in \Phi\}$. From (5) we see

(6) for any $A, B \in \Sigma$, one of the four sets
     $A \cap B$, $A^* \cap B$, $A \cap B^*$, $A^* \cap B^*$
     is finite.

Let $E = \{[A] \in \overline{P}(G) \mid A \in \Sigma\}$. As subset of $\overline{P}(G)$, E is partially ordered, and by the definition of Σ, E is closed under the actions of G and *.

(We remark that if H, or equivalently S, is infinite then the natural map $(\Sigma, \subseteq) \to (E, \leq)$ is bijective, so in this cas case one does not need the partial order $\leq$.)

It follows from (6) that for any $A, B \in \Sigma$, $[A]$ is comparable under $\leq$ to exactly one of $[B], [B^*]$, since B is proper. Hence for any $e, f \in E$, e is comparable to exactly one of f, f^*. Thus the construction of I.§9 applies to give a G-forest X with edge set E. To verify that X is a tree we must show that for any $e, f \in E$, the interval $[e, f]$ is finite, or equivalently the interval (e, f) is finite. Say $e = [A]$, $f = [C]$, $A, C \in \Sigma$, so $(e, f) = \{[B] \mid [A] < [B] < [C], B \in \Sigma\}$
$= \{g^1[B] \mid [A] < g^1[B] < [C], B \in \Phi, g \in G\}$.

But for any $B \in \Phi$ there are only finitely many such g since

(7) $\{g \in G \mid [A] < g^1[B] < [C]\}$
     $= \{g \in G \mid g[A] < [B]\} \cap \{g \in G \mid g[C^*] < [B^*]\} \stackrel{a}{\subseteq} B \cap B^*$

by 2.6 (i). Since Φ is finite, we deduce that $[e,f]$ is finite and hence X is a tree.

The stabilizer of each edge of X is finite by 2.6 (ii), and $G \backslash E$ is finite because Φ is finite, so the connected graph $G \backslash X$ is finite.

Thus the orders of the edge stabilizer subgroups are bounded, so for any subgroup K of G, I.8.8 says that K fixes a vertex of X if and only if every $G[e]$ is an almost right K-subset. But by I.9.1, $G[e] = \{g \in G \mid ge < e \text{ or } ge^* < e\}$, so if $e = [A]$, say, then $G[e] \stackrel{a}{=} A$ by 2.6 (i). Thus K fixes a vertex of X if and only if every $A \in \Sigma$ is an almost right K-subset, or equivalently, every $A \in \Phi$ is an almost right K-subset. In particular, H fixes a vertex of X, say v_H. Extend v_H to a (finite) connected transversal in X for the G-action. By the structure theorem, I.6.1, choosing a connecting family then gives a (finitary) decomposition of G over H, which can be seen to have all the desired properties. $\square$

In order to apply 3.2 inductively we need a rather combinatorial result. For any decomposition $(G_v \mid v \in V)$ of G we write, for each $v \in V$, $H_v = H \cap G_v$.

3.3 LEMMA. <u>Suppose</u> G <u>is finitely generated over</u> H <u>and let</u> $(G_v, q_e \mid v \in V, e \in E)$ <u>be a finitary decomposition of</u> G <u>over</u> H. <u>Then for each</u> $v \in V$, G_v <u>is finitely generated over</u> H_v, <u>and for each almost-right-invariant right</u> H-<u>subset</u> A <u>of</u> G, <u>there are only finitely many cosets</u> gG_v <u>that lie in neither</u> A <u>nor</u> A^*.

III   DECOMPOSITIONS

Proof. Take a set S that generates G finitely over H, and assume without loss of generality that $S - H \subseteq \{q_e | e \in E\} \cup \bigcup_V G_v$, and that for each $e \in E$, S contains G_e, $G_e^{q_e}$. For each $v \in V$ write $S_v = S \cap G_v$, and G_v' for the subgroup generated by S_v. By II.3.3, $G_v = G_v'$, so is finitely generated over H_v. Suppose gG_v is a coset that lies in neither A nor A^*. In $\Gamma(G, S_v)$, the subgraph $(gG_v)_{S_v}$ is connected so contains a path joining A to A^*, so meets δA. But δA is finite by 2.3, and the cosets gG_v are disjoint, so there can be at most $|\delta A|$ of them. □

We are now in a position to prove the main decomposition theorem.

3.4 THEOREM (Dunwoody [79]). <u>Suppose G is finitely generated over H. Let Φ be a finite set of almost-right-invariant right H-subsets of G. Then G has a finitary decomposition $(G_v | v \in V)$ over H such that for each $A \in \Phi$</u>,

(8) A <u>is an almost right G_v-subset for every</u> $v \in V$.

Proof. Let S generate G finitely over H. Then Φ is a finite set of S-cuts. By replacing the elements of Φ by their infinite connected components, we may assume that the elements of Φ are proper connected S-cuts (and right H-subsets), and by 2.8 we may further assume that Φ is a finite <u>full</u> set of proper connected S-cuts (and right H-subsets).

Take any finitary decomposition of G over H, $(G_v | v \in V)$, for example the trivial one vertex decomposition. If each

DECOMPOSITION THEOREMS §3

$A \in \Phi$ satisfies (8) then we are finished; if not, among the $A \in \Phi$ not satisfying (8) choose one that is minimal under $\underset{\sim}{\le}$.

Now consider any $v \in V$. By 3.3, there are only finitely many left cosets of G_v in G that lie in neither A nor A^*, say $g_1 G_v, \ldots, g_n G_v$. Now Φ is full, so for each $i, j = 1, \ldots, n$, and each $g \in G_v$, one of the four sets

(9) $\qquad g_i^{-1} A \cap g g_j^{-1} A, \quad g_i^{-1} A \cap g g_j^{-1} A^*, \quad g_i^{-1} A^* \cap g g_j^{-1} A, \quad g_i^{-1} A^* \cap g g_j^{-1} A^*$

has all its infinite components of the form xB, $B \in \Phi$, $B \underset{\sim}{\le} A$. By the minimality of A, each such B is an almost right G_v-set, so one of the four sets (9) is an almost right G_v-set, and in particular, has finite or cofinite intersection with G_v. Thus for any $B, C \in \Phi_v = \{g_i^{-1} A \cap G_v \mid i=1, \ldots, n\}$, and any $g \in G_v$, one of the four sets

$$B \cap gC, \quad B \cap gC^{*v}, \quad B^{*v} \cap gC, \quad B^{*v} \cap gC^{*v}$$

is finite, where $*v$ denotes the complement in G_v. In this situation, 3.2 gives a finitary decomposition of G_v over H_v, $(G_w \mid w \in W_v)$, such that each $B \in \Phi_v$ is an almost right G_w-subset for every $w \in W_v$. That is, each $A \cap g_i G_v$, $i = 1, \ldots, n$, is an almost right G_w-set for every $w \in W$. Hence so is

$$\left(A - \left(\bigcup_{i=1}^{n} g_i G_v \right) \right) \cup \bigcup_{i=1}^{n} (A \cap g_i G_v) = A.$$

The preceding applies to each $v \in V$, so we may expand all the G_v by 1.2 to get a finitary decomposition of G over H, $(g_w G_w g_w^{-1} \mid w \in \underset{V}{\cup} W_v)$, where the g_w are suitable elements of G. For each $v \in V$, each $B \in \Phi$ that is an almost right G_v-set is an

III   DECOMPOSITIONS

almost right G_w-subset for every $w \in W_v$; moreover, A is an almost right G_w-subset for every $w \in V W_v$. Since the elements of Φ are almost-right-invariant, any element of Φ that is an almost right G_w-subset is also an almost right $g_w G_w g_w^{-1}$-subset. So with the new finitary decomposition of G over H, we have increased the subset of elements of Φ satisfying (8). Since Φ is finite, continuing to expand in this way eventually gives a finitary decomposition of G over H with the desired property, and the theorem is proved. □

4.   THE RELATIONSHIP WITH DERIVATIONS

Let G be a group, H a subgroup of G, and R a nonzero ring.

Given any right G-module M, and a derivation $d: G \to M$, and a decomposition $(G_v \mid v \in V)$ of G, we shall say that the decomposition <u>admits</u> d if the restriction of d to each vertex group G_v is an inner derivation.

The main result of this chapter is that, for any projective right R[G]-module P, if $d: G \to P$ is a derivation such that G is finitely generated over the kernel of d, then G has a finitary decomposition which admits d.

We begin by fixing a substantial amount of notation.

4.1 DEFINITION. Let M be a right R[G]-module. In a natural way, the set D(G,M) of all derivations $d: G \to M$ is a right module over the ring $E = \text{End}_{\mathbb{Z}[G]} M$, and in particular over the subring $\text{End}_{R[G]} M$. Thus, for example, D(G,R[G]) is a <u>left</u>

THE RELATIONSHIP WITH DERIVATIONS §4

$R[G]$-module.

The map $ad: M \to D(G,M)$, $m \mapsto ad\ m$, is right E-linear, and has as image the E-submodule $InD(G,M)$ of $D(G,M)$ consisting of all inner derivations, and has kernel the set M^G of all elements of M fixed by G. For example, $InD(G,R[G]) = R[G]\delta_G$, where we write δ_G for $ad\ 1: G \to R[G]$. Since $R[G]^G = R[G]\sigma_G$, where

$$\sigma_G = \begin{cases} \sum_{g \in G} g & \text{if } G \text{ is finite,} \\ 0 & \text{if } G \text{ is infinite,} \end{cases}$$

we see $R[G]\delta_G \simeq R[G]/R[G]\sigma_G$.

A derivation $G \to M$ which vanishes on H is called an H-<u>derivation</u>. The set of all H-derivations $G \to M$ is an E-submodule of $D(G,M)$, and we shall denote it by $D_H(G,M)$. The intermediate E-submodule of all derivations $G \to M$ whose restriction to H is inner will be denoted $D_H^+(G,M)$. Notice $D_H^+(G,M) = D_H(G,M) + InD(G,M)$. □

For any set S we write $R[[S]]$ for the R-bimodule R^S; the element $(r_s)_S$, that is, $r_s \in R$ for all $s \in S$, will be written $\sum_S r_s \cdot s = \sum_S s \cdot r_s$. In a natural way, $R[S] \subseteq R[[S]]$.

We view $R[[G]]$ as $R[G]$-bimodule containing $R[G]$ as $R[G]$ sub-bimodule. An element $\alpha \in R[[G]]$ will be called <u>almost-right-invariant</u> if for all $g \in G$, we have $\alpha(1 - g) \in R[G]$. Let $\mathfrak{D}(R[[G]])$ denote the set of all almost-right-invariant elements of $R[[G]]$, clearly a left $R[G]$-submodule of $R[[G]]$. For any $\alpha \in \mathfrak{D}(R[[G]])$, the map $G \to R[G]$, $g \mapsto \alpha(1 - g)$, is clearly a derivation, and we shall denote it by $ad\ \alpha$, although it need not be inner. There is then a left $R[G]$-linear map

(10) $\qquad ad: \mathfrak{D}(R[[G]]) \to D(G, R[G])$.

III   DECOMPOSITIONS

The kernel of this map consists of all "constant" elements of
$R[[G]]$, that is, elements of the form $\sum_{g \in G} r.g$, $r \in R$. For any
subset A of G, we write χ_A for the element $\sum_{g \in A} g$ of
$R[[G]]$. Then the kernel of (10) is $R[G]\chi_G = R\chi_G$. We shall now
see that (10) is in fact surjective. For any $x \in G$, let
$\hat{x}: R[G] \to R$ be the coefficient-of-x map $\sum_{g \in G} r_g.g \mapsto r_x$.

4.2 PROPOSITION. <u>The left $R[G]$-linear map</u>

$$ad: \mathfrak{D}(R[[G]]) \to D(G, R[G])$$

<u>has an R-linear left inverse</u> $d \mapsto \sum_{x \in G} [xd]\hat{x}.x$, <u>and so induces
an $R[G]$-module isomorphism</u> $\mathfrak{D}(R[[G]])/R\chi_G \simeq D(G, R[G])$.

Proof. For any $d \in D(G, R[G])$, and any $g \in G$,

$$(\sum_{x \in G} [xd]\hat{x}.x)(1-g) = \sum_{x \in G} [xd]\hat{x}.x - \sum_{x \in G} [xd]\hat{x}.xg$$

$$= \sum_{x \in G} [(xg)d]\hat{xg}.xg - \sum_{x \in G} [xd.g]\hat{xg}.xg$$

$$= \sum_{x \in G} [gd]\hat{xg}.xg \quad \text{since } d \text{ is a derivation}$$

$$= gd. \quad \square$$

The preimage in $\mathfrak{D}(R[[G]])$ of $InD(G, R[G]) = R[G]\delta_G$ is
$R\chi_G + R[G]\chi_1$, the elements of finite or cofinite support. We
write $\mathfrak{D}_H(R[[G]])$ and $\mathfrak{D}_H^+(R[[G]])$ for the preimages in $\mathfrak{D}(R[[G]])$
of $D_H(G, R[G])$ and $D_H^+(G, R[G])$, respectively.

4.3 REMARKS. We think of the elements of $R[[G]]$ as formal sums
$\sum_{r \in R} r\chi_{A_r}$, the A_r pairwise disjoint and having union G. Such a
formal sum lies in $\mathfrak{D}_H(R[[G]])$ if and only if right multiplication
by any $s \in G$ changes only finitely many coefficients, and right

THE RELATIONSHIP WITH DERIVATIONS §4

multiplication by any $h \in H$ changes none of the coefficients; that is, $A_r h = A_r$ for all $r \in R$, $h \in H$, and for every $s \in G$, $A_r s \stackrel{a}{=} A_r$ for all $r \in R$, and $A_r s = A_r$ for almost all $r \in R$. For any generating set S of G over H, in $\Gamma(G,S)$,

(11) $$\bigcup_{r \in R} \delta A_r = \bigcup_{s \in S-H} (\bigcup_{r \in R} \{(g,gs) \mid g \in A_r - A_r s^{-1}\}).$$

Thus if $S - H$ is finite then so is (11), which means that only finitely many of the A_r are nonempty, in which case our formal sum is a genuine (finite) sum. □

4.4 PROPOSITION. *If G is finitely generated over H then $\mathcal{D}_H(R[[G]])$ is generated as left R-module by the elements of the form χ_A, A an almost-right-invariant right H-subset of G. Also, $\mathcal{D}_H^+(R[[G]]) = \mathcal{D}_H(R[[G]]) + R[G]\chi_1$.*

Proof. The first part follows from 4.3, and the second part from the fact that $D_H^+(G,R[G]) = D_H(G,R[G]) + R[G]\delta_G$. □

It is now a straightforward matter to translate 3.4 into a statement about derivations.

4.5 THEOREM (Dunwoody [79]). *Suppose that G is finitely generated over H. Then for any finite subset Ψ of $D_H^+(G,R[G])$, there exists a finitary decomposition of G over H that admits every element of Ψ.*

Proof. For each $d \in \Psi$ we may subtract a suitable inner derivation, and so assume that d vanishes on H. Thus we are taking $\Psi \subseteq D_H(G,R[G])$. By 4.2 and 4.4, there is a finite set Φ

III   DECOMPOSITIONS

of almost-right-invariant right H-subsets of G, such that for each $d \in \Phi$, $d = \text{ad}(\sum_{A \in \Phi} r_{A,d} \chi_A)$ for certain $r_{A,d} \in R$. By 3.4, there is a finitary decomposition $(G_v \mid v \in V)$ of G over H such that each element A of Φ is an almost right G_v-subset for every $v \in V$, from which it is easy to see that ad χ_A lies in $D_{G_v}^+(G, R[G])$ for every $v \in V$, and hence the same is true of each element of Ψ. □

For Chapter IV we need the following formulation.

4.6 THEOREM. <u>If P is a projective right R[G]-module, and d:G → P is a derivation such that G is finitely generated over the kernel of d, then there exists a finitary decomposition of G admitting d, and having ker d as one of the vertex groups.</u>

Proof. There is no harm in adding an R[G]-summand to P, so we may assume that P is a free R[G]-module. Since G is finitely generated over ker d, the image of d lies in a finitely generated submodule of P, so we may assume that P has finite rank. Now 4.5 gives a finitary decomposition of G over ker d admitting each component of d, so admitting d. Now consider the vertex group G_v containing ker d. There exists an element p of P such that ad p agrees with d on G_v. In particular, p is fixed by ker d. If ker d is infinite then p = 0, so G_v = ker d. If ker d is finite, then by adjoining a new edge and a new vertex to the decomposition, we may assume that ker d appears as one of the vertex groups. □

4.7 REMARKS. Let us examine what 4.5 says in terms of groups

THE RELATIONSHIP WITH DERIVATIONS §4

acting on trees. Thus, let $(G_v, q_e \mid v \in V(Y), e \in E(Y))$ be any decomposition of G, and let X be the corresponding standard tree, cf I.4.4. Then G has a left action on X which we convert into a right action by setting $x.g = g^{-1}.x$ for all $x \in X$, $g \in G$. We shall view

$$E(X) = \bigvee_{e \in E(Y)} G_e \backslash G, \qquad V(X) = \bigvee_{v \in V(Y)} G_v \backslash G,$$

so the edges of X have the form

$$G_{\iota e}g \quad\underline{\qquad}\quad G_e g \quad\underline{\qquad}\quad G_{\tau e}q_e^{-1}g \quad .$$

For each $y \in Y$, we write $\hat{y}$ for the natural representative in X, that is, $G_y 1$. Then for any $v_0 \in V(Y)$ there is a G_{v_0}-derivation $X(\hat{v}_0, \hat{v}_0\_): G \to R[E(X)]$, $g \mapsto X(\hat{v}_0, \hat{v}_0 g)$, cf I.§1, I.§5. Our original decomposition admits this derivation, and we shall see later that any other G_{v_0}-derivation $d: G \to M$ to a right $R[G]$ module M, factors through $X(\hat{v}_0, \hat{v}_0\_)$ if and only if our original decomposition admits d. □

4.8 DEFINITION. For any right G-set U, there is a right $R[G]$-linear map

$$\varepsilon_U : R[U] \to R, \qquad \sum_{u \in U} r_u . u \mapsto \sum_{u \in U} r_u,$$

called the <u>augmentation map</u>, and the kernel is denoted $\omega_R(U)$, and is called the <u>augmentation module</u> of U over R. □

III   DECOMPOSITIONS

4.9 PROPOSITION. <u>Let M be a right $R[G]$-module, U a right
G-set, and z an element of U. The G_z-derivation
$d_z: G \to \omega_R(U)$, $g \mapsto z - zg$, induces a right $\mathrm{End}_{R[G]}M$-linear map</u>

$$\mathrm{Hom}_{R[G]}(\omega_R(U),M) \to D(G,M), \quad \alpha \mapsto d_z \cdot \alpha,$$

<u>and the image of this map is</u> $D_{G_z}(G,M) \cap \bigcap_{u \in U} D^+_{G_u}(G,M)$.

<u>In particular, there is a surjective right</u> $\mathrm{End}_{R[G]}M$-<u>linear
map</u> $\mathrm{InD}(G,M) \oplus \mathrm{Hom}_{R[G]}(\omega_R(U),M) \to \bigcap_{u \in U} D^+_{G_u}(G,M)$.

Proof. Since d_z agrees with $\mathrm{ad}(z - u)$ on G_u for each $u \in U$,
we see that for any $\alpha: \omega_R(U) \to M$, $d_z \cdot \alpha$ vanishes on G_z and is
inner on each G_u. Conversely, let $d: G \to M$ be a G_z-derivation
that is inner on each G_u. For each $u \in U$ there exists $m_u \in M$
such that d agrees with $\mathrm{ad}\, m_u$ on G_u, and the m_u may be
chosen so that $m_z = 0$ and for each $u \in U$, and $g \in G$,
$m_{ug} = m_u g + gd$. (To see this, extend z to a transversal S
in U for the G-action, and for each $s \in S$, choose $m_s \in M$ so
that d agrees with $\mathrm{ad}\, m_s$ on G_s, and $m_z = 0$, and then define
$m_{sg} = m_s g + gd$ for each $g \in G$.) Thus, there is an R-linear
map $\alpha: \omega_R(U) \to M$, $(\sum_{u \in U} u \cdot r_u) \mapsto (\sum_{u \in U} m_u \cdot r_u)$. For any $g \in G$,
$\sum_{u \in U} m_u g r_u = \sum_{u \in U} (m_{ug} - gd) r_u = \sum_{u \in U} m_{ug} r_u$ if $\sum_{u \in U} r_u = 0$, so
α is right $R[G]$-linear. Further, for any $g \in G$,
$(g)d_z \cdot \alpha = (z - zg)\alpha = -m_z + m_{zg} = gd$ since $m_z = 0$, so d
is in the image of our map, as desired. □

(We remark that the foregoing amounts to the exactness at one
term of a long exact Ext sequence.)

THE RELATIONSHIP WITH DERIVATIONS   §4

4.10 COROLLARY. <u>For any right</u> $R[G]$-<u>module</u> M,
$\text{Hom}_{R[G]}(\omega_R(H\backslash G),M) \simeq D_H(G,M)$ <u>as right</u> $\text{End}_{R[G]}M$-<u>modules</u>. □

4.11 REMARK. $R[H\backslash G] = \underset{Hg \in H\backslash G}{\oplus} R[H\backslash Hg] = \underset{H\backslash G}{\oplus} R \otimes_{R[H]} R[Hg]$
$= R \otimes_{R[H]} R[G]$. Thus, for any right $R[G]$-module M,
$\text{Hom}_{R[G]}(R[H\backslash G],M) = \text{Hom}_{R[H]}(R,M) \simeq M^H$, as right $\text{End}_{R[G]}M$ modules. □

4.12 PROPOSITION. <u>Let</u> M <u>be a right</u> $R[G]$-<u>module, and</u>
$(G_v, q_e \mid v \in V(Y), e \in E(Y))$ <u>be any decomposition of</u> G. <u>Write</u>
X <u>for the corresponding standard right</u> G-<u>tree, and choose any</u>
$v_0 \in V(Y)$. <u>Then there is a surjective right</u> $\text{End}_{R[G]}M$-<u>linear map</u>

(12) $\text{Hom}_{R[G]}(R[E(X)],M) \to D_{G_{v_0}}(G,M) \cap \underset{v \in V(Y)}{\cap} D^+_{G_v}(G,M)$

$\alpha \mapsto X(\hat{v}_0, \hat{v}_0\_) \cdot \alpha.$

<u>In particular, there is a surjective right</u> $\text{End}_{R[G]}M$-<u>linear map</u>
$\text{InD}(G,M) \oplus \underset{e \in E(Y)}{\Pi} M^{G_e} \to \underset{v \in V(Y)}{\cap} D^+_{G_v}(G,M).$

Proof. From I.1.1, we have an isomorphism $R[E(X)] \simeq \omega_R(V(X))$,
$e \mapsto \iota e - \tau e$, of right $R[G]$-modules. In particular,
$X(\hat{v}_0, \hat{v}_0 g) \mapsto \hat{v}_0 - \hat{v}_0 g$, so the result follows from 4.9.

The second part follows from 4.11. □

In light of 4.12, 4.5 may be viewed as saying that there exists a right G-tree X with finite edge stabilizers, and with X/G finite, such that each element of Ψ arises, modulo inner derivations, from an $R[G]$-linear map $R[E(X)] \to R[G]$.

It is interesting to examine these derivations in greater detail.

III   DECOMPOSITIONS

Thus, let X be a right G-tree, and let e be any edge of X, and suppose G_e is finite. Let $\alpha: R[E(X)] \to R[G]$ be the $R[G]$ linear map which sends e to σ_{G_e} and vanishes on all other orbits in $E(X)$. For any vertex v of X, we have a derivation $d: G \to R[G]$, $g \mapsto X(v, vg)\alpha$. Now consider the preimage of d in $R[[G]]$. For each $x \in G$,

$$[xd]\hat{x} = \begin{cases} +1 & \text{if } v \to ex \to vx, \text{ or equivalently, } vx^{-1} \to e \to v \\ -1 & \text{if } vx \to ex \to v, \text{ or equivalently, } v \to e \to vx^{-1} \\ 0 & \text{otherwise.} \end{cases}$$

To conform with the notation used in I.§8, we write $G[e, v] = \{x \in G \mid e \to vx^{-1}\}$. By considering the cases $e \to v$, $v \to e$, separately, we see from 4.2 that $d = -\text{ad } \chi_{G[e,v]}$.

5. ACCESSIBILITY

Let G be a group, H a subgroup of G, and R a nonzero ring.

This section, which is loosely based on Dunwoody [79], considers various consequences of 4.5. None of the results proved in this section will be used in Chapter IV.

A derivation is called <u>outer</u> if it is not inner.

A decomposition of G is called <u>proper</u> if G does not occur as one of the vertex groups.

An almost-right-invariant subset A of G is called <u>proper</u> if both A and A^* are infinite.

These concepts are related as follows.

5.1 THEOREM. The following four statements are equivalent.
(a) There exists an action of G on a tree X with finite edge stabilizers, such that G fixes no vertex of X, and H fixes some vertex of X.
(a') There exists a proper decomposition of G over H with finite edge groups.
(b) There exists an action of G on a tree X with finite edge stabilizers, such that G fixes no vertex of X, and H fixes some vertex v_0 of X, and one of the following holds:
(13)   X/G has precisely one edge.
(14)   There is an "infinite path" $v_0, e_1^{\varepsilon_1}, v_1, \ldots$ in X which is a transversal for the G-action, and such that $H \subset G_{e_1} \subset G_{e_2} \subset \ldots$ is a strictly ascending chain (of finite groups, whose union is all of G).
(b') At least one of the following holds.
(15) $G = A \amalg_C B$, C finite, $H \le A$, $A > C < B$.
(16) $G = HNN<\alpha, \beta : C \to A; t>$, C finite, $H \le A$.
(17)   G is infinite locally finite, and H is finite.

Further, these imply the following two statements which are equivalent to each other.
(c) There exists an outer H-derivation $G \to R[G]$.
(c') There exists a proper almost-right-invariant right H-subset of G.

If G is finitely generated over H then all six statements are equivalent (and here (14),(17) do not occur).

Proof. (a)⇔(a'), (b)⇔(b') follow from the structure theorem and its converse, I.6.1, I.5.3.
(c)⇔(c') by the remarks made in 4.3; for if $d: G \to R[G]$ is

III   DECOMPOSITIONS

an outer derivation then $d = \text{ad}(\sum_{r \in R} r\chi_{A_r})$, where the A_r are pairwise disjoint almost-right-invariant right H-subsets of G, such that the union is all of G and the union of any subfamily of these is again an almost-right-invariant right H-subset. It is an easy matter to choose a subfamily whose union is proper. Conversely, if A is a proper almost-right-invariant right H-subset of G then $\text{ad } \chi_A : G \to R[G]$ is an outer H-derivation.

(a)⇔(b). Clearly (b)⇒(a). To see the converse, observe that for any G-subset $\bar{E}$ of E(X) we can construct the graph $\bar{X}$ having $E(\bar{X}) = \bar{E}$ and $V(\bar{X})$ equal to the set of connected components of $X - \bar{E}$. It can be seen that $\bar{X}$ is a tree with a natural G-action extending the G-action on $\bar{E}$. (This is an interesting example to illustrate the construction of I.§9.) We now consider two cases: either some element g of G shifts an edge e of X, or no element of G shifts an edge of X. In the former case we take $\bar{E} = eG$, and then $\bar{X}$ is as in (13) (since by I.8.3 no vertex of $\bar{X}$ is fixed by g); in the latter case, I.8.4 shows there exists an "infinite path" $v_0, e_1^{\varepsilon_1}, v_1, \ldots$ with $H \subseteq G_{v_0} = G_{e_1} \subseteq G_{v_1} = G_{e_2} \subseteq \ldots$ an ascending chain of (finite) subgroups of G whose union is all of G. We choose a subsequence $e_{i_1}, e_{i_2}, \ldots$ such that $H \subset G_{e_{i_1}} \subset G_{e_{i_2}} \subset \ldots$ and take $\bar{E}$ to be the G-subset of E(X) it generates. Then $\bar{X}$ is as in (14). Thus (b) holds.

(b)⇒(c'). If (14) holds then $\cup_{n \geq 1} (G_{e_{2n}} - G_{e_{2n-1}})$ is a proper almost-right-invariant right H-subset of G. If (13) holds then, since G does not stabilize any vertex of X, I.8.4 shows there is an edge e shifted by an element of G, and then

ACCESSIBILITY §5

$G[e]$ is an almost-right-invariant almost right H-subset of G by I.8.6, and is proper by I.8.7. It follows that (c') holds.

Finally, if G is finitely generated over H then (c) ⇒ (a') by 4.5. (Alternatively, one can use 3.2 and 2.8, which is more in the spirit of Stallings [68] proof of (c') ⇒ (b') for $H = 1$.) □

Suppose that G is finitely generated over H. We shall say that G is (finitarily) <u>decomposable over</u> H if the above six (equivalent) conditions are satisfied, that is, G has a proper finitary decomposition over H; otherwise G is (finitarily) <u>indecomposable over</u> H. (We shall not be using the qualifier "finitarily" anywhere since we shall be dealing with no other type of decomposability.) For example, if G is finite then G is indecomposable over H by II.3.5 or 4.2. We shall usually omit the qualifier "over H" if $H = 1$.

Since $D_H^+(G,R[G]) = D_H(G,R[G]) + R[G]\delta_G$, we see that 5.1 (c) fails if and only if $D_H^+(G,R[G]) = R[G]\delta_G$, so the latter is a criterion for indecomposability.

We shall need a slightly more general version of this.

Since $R[H]$ is a subring of $R[G]$, we may view $D(H,R[H])$ as a subset of $D(H,R[G])$. (In particular, $\delta_H : H \to R[H]$, $h \mapsto 1 - h$, is thus an element of $D(H,R[G])$.)

III   DECOMPOSITIONS

5.2 LEMMA. *Let K be a subgroup of G containing H. The embedding $D_H^+(K,R[K]) \subseteq D_H^+(K,R[G])$ induces an injective $R[G]$ linear map which will be treated as an embedding*
$$R[G] \otimes_{R[K]} D_H^+(K,R[K]) \subseteq D_H^+(K,R[G]).$$

If K is finitely generated over H then equality holds.

Proof. $R[G] \otimes_{R[K]} D_H^+(K,R[K]) = \bigoplus_{gK \in G/K} gR[K] \otimes_{R[K]} D_H^+(K,R[K])$

$= \bigoplus_{gK \in G/K} D_H^+(K,gR[K]) \subseteq D_H^+(K, \bigoplus_{gK \in G/K} gR[K]) = D_H^+(K,R[G]).$

Equality holds if and only if each derivation $d: K \to R[K]$ that is inner on H has image lying in a finite sum of $gR[K]$, and in fact it suffices to consider only H-derivations d. Suppose S generates K finitely over H. Then for any H-derivation $d: K \to R[G]$, we have $(K)d \subseteq \sum_{s \in S} (s)d.R[G]$, which completes the proof. □

5.3 INDECOMPOSABILITY CRITERIA. *Let K be a subgroup of G containing H and finitely generated over H. Then the following are equivalent.*
(a) K *is indecomposable over* H.
(b) $D_H^+(K,R[K]) = R[K]\delta_K \quad (\simeq R[K]/R[K]\sigma_K)$.
(c) $D_H^+(K,R[G]) = R[G]\delta_K \quad (\simeq R[G]/R[G]\sigma_K)$.
(d) $D_H(K,R[G]) = R[G]\sigma_H\delta_K \quad (\simeq R[G]\sigma_H/R[G]\sigma_K)$.

Proof. (a)$\Longleftrightarrow$(b) by the remark preceding 5.2.
(b)$\Longleftrightarrow$(c) since by 5.2,
$$D_H^+(K,R[G])/R[G]\delta_K = R[G] \otimes_{R[K]} (D_H^+(K,R[K])/R[K]\delta_K).$$
(c)$\Rightarrow$(d) since $D_H(K,R[G]) \cap R[G]\delta_K = R[G]\sigma_H\delta_K$.
(d)$\Rightarrow$(c) since $D_H^+(K,R[G]) = D_H(K,R[G]) + R[G]\delta_K$. □

ACCESSIBILITY §5

Let us record one consequence of 5.3.

5.4 COROLLARY. <u>If K is a subgroup of G that contains H, and is finitely generated over H, and indecomposable over H, then</u> $D_K^+(G,R[G]) = D_H^+(G,R[G])$.

<u>In particular, if K is finitely generated and indecomposable then</u> $D_K^+(G,R[G]) = D(G,R[G])$. □

Suppose that G is finitely generated over H. Then G is said to be <u>accessible over</u> H if G has a finitary decomposition $(G_v \mid v \in V)$ over H, such that, for each $v \in V$, G_v is indecomposable over H_v; that is, a sort of "atomic" decomposition. (Here, since the edge groups are finite, each H_v is finite for $v \neq v_H$.) If G is not accessible over H then it is said to be <u>inaccessible over</u> H. As usual, we omit the qualifier "over H" if $H = 1$.

No example is known where G is finitely generated over H and known to be inaccessible over H. Wall [71] conjectured that every finitely generated group is accessible, but so far the problem remains open.

We now turn our attention to proving Dunwoody's results which relate accessibility to derivations.

III DECOMPOSITIONS

5.5 LEMMA (Bamford-Dunwoody [76]). <u>Let</u> $(G_v, q_e | v \in V(Y), e \in E(Y))$ <u>be a decomposition of</u> G <u>over</u> H, <u>and let</u> $v_0 \in V(Y)$. <u>If for each edge</u> e <u>in</u> $star(v_0)$, G_e <u>is finitely generated and indecomposable then the restriction map</u> $D_H^+(G, R[G]) \to D_{H_{v_0}}^+(G_{v_0}, R[G])$ <u>is surjective</u>.

Proof. Let $d_{v_0}: G_{v_0} \to R[G]$ be any derivation. Write T for the maximal subtree of Y. For each edge e of T lying in $star(v_0)$, the restriction of d_{v_0} to G_e is inner by 4.9 (c), say it equals ad $\tilde{e}$, $\tilde{e} \in R[G]$. For each vertex $v \neq v_0$, define $d_v: G_v \to R[G]$ to be ad $\tilde{e}_1$ where $e_1^{\varepsilon_1}, \ldots, e_n^{\varepsilon_n}$ is the geodesic in T from v_0 to v. For each edge e of T, $d_{\iota e}$ and $d_{\tau e}$ clearly agree on G_e. Let e be any edge of Y, and consider the derivation $d_e: G_e \to R[G]$, $g \mapsto (g^{q_e}) d_{\tau e} q_e^{-1} - (g) d_{\iota e}$. If $e \notin star(v_0)$ then $d_{\iota e}$ and $d_{\tau e}$ are both inner and hence so is d_e; if $e \in star(v_0)$ then d_e is inner by 5.3 (c). Thus for each $e \in E(Y)$ we may choose $m_e \in R[G]$ such that ad $m_e q_e^{-1}$ agrees with d_e on G_e; further, for the edges e of T we may take $m_e = 0$. By I.5.2 there is a derivation $d: G \to R[G]$ such that d agrees with d_v on each G_v, and sends each q_e to m_e ($v \in V(Y)$, $e \in E(Y)$). If $v_H \neq v_0$ then d_{v_H} is inner on G_{v_H} so d is inner on H; if $v_H = v_0$ and d_{v_0} is inner on H then d is inner on H. This proves the lemma. □

We can now prove our first characterization of accessibility.

5.6 THEOREM (Dunwoody [79]). <u>Suppose</u> G <u>is finitely generated over</u> H. <u>Then</u> G <u>is accessible over</u> H <u>if and only if</u> $D_H^+(G, R[G])$ <u>is finitely generated as left</u> $R[G]$-<u>module</u>.

Proof. ⇒ Suppose G is accessible over H, say $(G_v, q_e \mid v \in V, e \in E)$ is a finitary decomposition of G over H such that for each $v \in V$, G_v is indecomposable over H_v. Choose any $v_0 \in V$ and apply the last part of 4.12 to get a surjective left $R[G]$-linear map from

$$R[G]\delta_G \oplus \prod_{e \in E} R[G]^{G_e}$$
$$= R[G]\delta_G \oplus \bigoplus_{e \in E} R[G]\sigma_{G_e}$$

onto

$$\bigcap_V D^+_{G_v}(G, R[G])$$
$$= \bigcap_V D^+_{H_v}(G, R[G]) \quad \text{by 5.4}$$
$$= D^+_H(G, R[G]).$$

This gives $|E| + 1$ elements that generate $D^+_H(G, R[G])$.

⇐ Suppose $D^+_H(G, R[G])$ is generated as left $R[G]$-module by some finite set Ψ. By 4.5 there is a finitary decomposition $(G_v \mid v \in V)$ of G over H such that for each $v \in V$ the restriction map

(18) $\qquad D^+_H(G, R[G]) \to D^+_{H_v}(G_v, R[G])$

carries Ψ to $R[G]\delta_{G_v}$. But by 5.5, (18) is surjective, so the image of Ψ is a generating set, and therefore $D^+_{H_v}(G_v, R[G]) = R[G]\delta_{G_v}$. Now 5.3 shows G_v is indecomposable over H_v, which proves that G is accessible over H. □

III   DECOMPOSITIONS

5.7 COROLLARY. <u>Suppose</u> G <u>is finitely generated over</u> H. <u>Then</u> G <u>is accessible over</u> H <u>if and only if</u> $D_H(G,R[G])$ <u>is finitely generated as left</u> $R[G]$-<u>module</u>.

Proof. Recall that $D_H^+(G,R[G]) = D_H(G,R[G]) + R[G]\delta_G$ and $D_H(G,R[G]) \cap R[G]\delta_G = R[G]\sigma_H\delta_G$. Thus

$$D_H^+(G,R[G])/R[G]\delta_G \simeq D_H(G,R[G])/R[G]\sigma_H\delta_G$$

so $D_H^+(G,R[G])$ is finitely generated as left $R[G]$-module if and only if $D_H(G,R[G])$ is. □

With this result, the generality of R being arbitrary ceases to be important, and we turn to the case $R = \mathbb{Z}$ and study the abelian group

$$D_H^+(G) = \mathbb{Z} \otimes_{\mathbb{Z}[G]} D_H^+(G,\mathbb{Z}[G])$$

where $\mathbb{Z}$ is viewed as right $\mathbb{Z}[G]$-module with trivial G-action. This abelian group was studied in Bamford-Dunwoody [76] in the case $H = 1$, and the relative version was introduced in Dunwoody [79], where it was called $A(G,H)$.

We remark that if H is finitely generated indecomposable then by 5.4, $D_H^+(G) = D_1^+(G)$.

5.8 PROPOSITION. <u>If</u> G <u>is finitely generated over</u> H <u>and indecomposable over</u> H <u>then</u>

$$D_H^+(G) \simeq \begin{cases} \mathbb{Z} & \underline{if}\ G\ \underline{is\ infinite} \\ \mathbb{Z}_{|G|} & \underline{if}\ G\ \underline{is\ finite}. \end{cases}$$

Proof. By 5.3, $D_H^+(G,\mathbb{Z}[G]) = \mathbb{Z}[G]\delta_G \simeq \mathbb{Z}[G]/\mathbb{Z}[G]\sigma_G$, so we have a presentation

$$Z[G] \xrightarrow{\sigma_G} Z[G] \longrightarrow D_H^+(G, Z[G]) \longrightarrow 0.$$

Applying $Z \otimes_{Z[G]} -$ then gives a presentation

$$Z \xrightarrow{n_G} Z \longrightarrow D_H^+(G) \longrightarrow 0$$

where

$$n_G = \begin{cases} 0 & \text{if } G \text{ is infinite} \\ |G| & \text{if } G \text{ is finite.} \end{cases} \quad \square$$

Recall that there is an exact sequence

$$0 \to \omega \to Z[G] \xrightarrow{\varepsilon} Z \to 0$$

where $(\Sigma\, n_g g)\varepsilon = \Sigma\, n_g$, and ω is called the <u>augmentation ideal</u> of G. Applying $- \otimes_{Z[G]} D_H^+(G, Z[G])$ gives a presentation

$$0 \to \omega D_H^+(G, Z[G]) \longrightarrow D_H^+(G, Z[G]) \xrightarrow{\overline{}} D_H^+(G) \to 0$$

where $\overline{d} = 1 \otimes d$. Thus if we are given a d_0 such that $\overline{d}_0 = 0$ then we can write $d_0 = \sum_{i=1}^{n} w_i d_i$, $w_i \in \omega$, $d_i \in D_H^+(G, Z[G])$. If G is finitely generated over H then by 4.5 there is a finitary decomposition $(G_v \mid v \in V)$ of G over H admitting $d_0, \ldots, d_n$, that is, there exist $x_{i,v} \in Z[G]$ ($i = 0, \ldots, n$; $v \in V$) such that for each i, v, d_i agrees with $\text{ad}\, x_{i,v}$ on G_v. So for each v, $\text{ad}\, x_{0,v} = \sum_{i=1}^{n} \text{ad}\, w_i x_{i,v}$ on G_v, that is, $x_{0,v} - \sum_{i=1}^{n} w_i x_{i,v} \in Z[G]\sigma_{G_v}$. In particular, either G_v is finite or $(x_{0,v})\varepsilon = 0$. We record a special case of this.

III DECOMPOSITIONS

5.9 PROPOSITION (Dunwoody [79]). <u>Suppose G is finitely generated over H. If there is a positive integer m such that $m\bar{\delta}_G = 0$ in $D_H^+(G)$ then G has a finitary decomposition over H in which all the vertex groups are finite.</u>

Proof. We are in the situation of the remarks preceding the proposition, with $d_0 = \mathrm{ad}\, m$. Thus we may take $x_{0,v} = m$ for each v, and since $(m)\varepsilon = m \neq 0$, each G_v is finite. □

(Of course it follows that in the above situation H is finite and G is accessible.)

In order to use the abelian group $D_H^+(G)$ to measure the accessibility of G over H we want some way of comparing $D_H(G)$ with the various $D_{H_v}(G_v)$ that arise from a decomposition. Recall that for any subgroup K of G there is a natural map

$$D_H^+(G, \mathbb{Z}[G]) \to D_{H \cap K}^+(K, \mathbb{Z}[G]) \supseteq \mathbb{Z}[G] \otimes_{\mathbb{Z}[G]} D_{H \cap K}^+(K, \mathbb{Z}[K])$$

where the latter containment is equality if K is finitely generated over $H \cap K$. In this case, tensoring with $\mathbb{Z}$ induces a canonical map $D_H^+(G) \to D_{H \cap K}^+(K)$.

5.10 PROPOSITION (Dunwoody [79]). <u>Suppose G is finitely generated over H, and let $(G_v, q_e \mid v \in V(Y), e \in E(Y))$ be a finitary decomposition of G over H. Then for any $v_0 \in V(Y)$, if $G_{v_0} \subset G$ then the canonical map $D_H^+(G) \to D_{H_{v_0}}^+(G_{v_0})$ is</u> surjective and has nonzero kernel.

Proof. From 5.5 we know that $D_H^+(G, \mathbb{Z}[G]) \to D_{H_{v_0}}^+(G_{v_0}, \mathbb{Z}[G])$ is surjective, and tensoring with $\mathbb{Z}$ then shows that $D_H^+(G) \to D_{H_{v_0}}^+(G_{v_0})$ is surjective. It remains to find an element in the kernel that is not zero. We may assume without loss of generality that for each $v \in V(Y) - \{v_0\}$, G_v is either indecomposable over H_v or is inaccessible over H_v. (For if some G_v is decomposable over H_v and accessible over H_v then we may expand the decomposition by 1.2 to get a new finitary decomposition of G over H but having some conjugate $G_{v_0}^g$ of G_{v_0} as a vertex group, where $g = 1$ if $v_0 = v_H$, and otherwise $H \cap G_{v_0}$, $H \cap G_{v_0}^g$ are both finite. Thus it can be seen that there is a commuting diagram

$$\begin{array}{ccc} D_H^+(G,\mathbb{Z}[G]) & \longrightarrow & D_{H \cap G_{v_0}}^+(G_{v_0},\mathbb{Z}[G]) \\ \Vert & & \Vert \\ D_{H^g}^+(G^g,\mathbb{Z}[G]) & \longrightarrow & D_{H^g \cap G_{v_0}^g}^+(G_{v_0}^g,\mathbb{Z}[G]) \\ \Vert & & \Vert \\ D_H^+(G,\mathbb{Z}[G]) & \longrightarrow & D_{H \cap G_{v_0}^g}^+(G_{v_0}^g,\mathbb{Z}[G]) \end{array}$$

so the new decomposition can be used in place of the old.)

Let X be the standard tree associated to the decomposition, cf 4.7. For each edge e of Y, let $\alpha_e : R[E(X)] \to R[G]$ be the right $R[G]$-linear mapping sending $\hat{e}$ to σ_{G_e} and vanishing on all other orbits in $E(X)$, and let $d_e : G \to R[G]$ be the derivation $X(\hat{v}_0, \hat{v}_{0-})\alpha_e$. Then d_e is inner on G_{v_H} so we have an element $\bar{d}_e$ of $D_H^+(G)$, and further, since d_e vanishes on G_{v_0}, $\bar{d}_e$ is in the kernel of $D_H^+(G) \to D_{H_{v_0}}^+(G_{v_0})$. It remains to choose e in such a way that $\bar{d}_e \neq 0$.

If Y is not a tree, then there is some e with $q_e \neq 1$, and then $(q_e^{-1})d_e = (X(\hat{v}_0, \hat{\tau e}) + \hat{e} + X(\hat{\tau e}, \hat{v}_0)q_e^{-1})\alpha_e = \sigma_{G_e} \notin \omega$ so $d_e \notin D_H^+(G, \omega) \supseteq \omega D_H^+(G,\mathbb{Z}[G])$, and thus $\bar{d}_e \neq 0$ for this choice of e.

III   DECOMPOSITIONS

This leaves the case where Y is a tree. Since $G_{v_0} \not\leq G$, there must be some vertex v of Y such that $G_{v_0} \not\leq G_v$. Thus, in the geodesic $e_1^{\varepsilon_1}, \ldots, e_n^{\varepsilon_n}$ from v_0 to v there is some term e^ε such that $G_e \not\leq G_{\tau e}\varepsilon$ and we may choose v so that $v = \tau e^\varepsilon$. We shall show that for this choice of e, $\bar{d}_e \neq 0$ in $D_H^+(G)$. The restriction of $X(\hat{v}_0, \hat{v}_{0-})$ to G_v is ad $X(\hat{v}_0, \hat{v})$, so the restriction of d_e to G_v is ad $\varepsilon\sigma_{G_e}$; thus, under $D_H^+(G) \to D_{H_v}^+(G_v)$, $\bar{d}_e$ maps to

(19) $\varepsilon\overline{\sigma_{G_e}\delta_{G_v}} = \varepsilon|G_e|\overline{\delta}_{G_v}$ in $D_{H_v}^+(G_v)$.

Suppose this is zero; we get a contradiction as follows. By 5.9, G_v is accessible over H_v, so by our choice of decomposition, G_v is indecomposable over H_v, so by 5.9 again, G_v is finite, and by 5.8, $\bar{\delta}_{G_v}$ has order $|G_v|$ in $D_{H_v}^+(G_v)$. But $G_e \subset G_v$, so $|G_e| < |G_v|$, so $|G_e|\bar{\delta}_{G_v} \neq 0$. Hence (19) is nonzero, and thus $\bar{d}_e \neq 0$, as claimed. □

5.11 DEFINITION. For any finitely generated abelian group A, let us define the <u>size</u> of A, size(A), to be the ordered pair (rank(A/tA), |tA|), where tA is the torsion subgroup of A. Two such pairs will be compared lexicographically, that is, $(a_1, a_2) > (b_1, b_2)$ if $a_1 > b_1$ or if $a_1 = b_1$, $a_2 > b_2$. (This defines a well-ordering of the set of pairs of nonnegative integers.)

Suppose G is finitely generated over H and that the abelian group $D_H^+(G)$ is finitely generated. For any finitary decomposition $D = (G_v | v \in V)$ of G over H we can then

ACCESSIBILITY §5

define the <u>defect</u> of D, defect(D), to be the finite descending sequence

$$\text{size}(D_H^+(G_{v_1})), \quad \text{size}(D_H^+(G_{v_2})), \quad \ldots, \quad \text{size}(D_H^+(G_{v_n}))$$

where $v_1, v_2, \ldots, v_n$ is an enumeration of those elements v of V such that G_v is <u>decomposable</u> over H_v, and the enumeration is such that

$$\text{size}(D_H^+(G_{v_1})) \geq \text{size}(D_H^+(G_{v_2})) \geq \ldots \geq \text{size}(D_H^+(G_{v_n})).$$

(For example, if G is decomposable over H but D is not a proper decomposition then $n = 1$, and defect(D) consists of one term, $\text{size}(D_H^+(G))$.)

Two such sequences will be compared lexicographically, that is, $(p_1, \ldots, p_n) > (q_1, \ldots, q_m)$ if

there exists $i < n$ such that $p_1 = q_1, \ldots, p_i = q_i$, $p_{i+1} > q_{i+1}$,

or

$n > m$ and $p_1 = q_1, \ldots, p_m = q_m$.

(Again, this is a well-ordering.)

Suppose now that we expand D as in 1.2 to get a new finitary decomposition $D' = (g_w G_w g_w^{-1} \mid w \in \vee_V W_v)$ of G over H. Here we shall say that D' is an <u>expansion</u> of D, although to be precise one should say that D' <u>can be obtained by</u> an expansion of D. Let us see how the defects of D and D' compare. Let $v \in V$. If the corresponding decomposition $D_v = (G_w \mid w \in W_v)$ of G_v over H_v is <u>not</u> proper then the contribution of v to defect(D)

III   DECOMPOSITIONS

is the same as the total contribution of the $w \in W_v$ to defect(D'), namely, size($D_{H_v}^+(G_v)$) if G_v is decomposable over H_v, and nothing if G_v is indecomposable over H_v. If D_v is a proper decomposition of G_v over H_v then v contributes exactly one term, size($D_{H_v}^+(G_v)$), to defect(D), while any terms contributed to defect(D') by the $w \in W_v$ are smaller than this, for by 5.10, the natural map

(20) $\qquad D_{H_v}^+(G_v) \to D_{H_w}^+(G_w) \simeq D_{H \cap g_w G_w g_w^{-1}}^+(g_w G_w g_w^{-1})$

is surjective and has nonzero kernel, so the size decreases. (To see the isomorphism step in (20), recall that either $w = w_H$ and $g_w = 1$, and we have equality, or $w \neq w_H$ so H_w and $H \cap g_w G_w g_w^{-1}$ are both finite, so may be replaced with 1 in (20).)

Let us say that D' is a <u>proper</u> expansion of D if for some v the decomposition D_v is proper; and otherwise D' is an <u>improper</u> expansion of D. Thus we have shown that if D' is an improper expansion of D then defect(D) = defect(D'), while if D' is a proper expansion of D then defect(D) > defect(D'). □

5.12 THEOREM (Dunwoody [79]). <u>If G is finitely generated over H then the following are equivalent</u>.
(a) $D_H^+(G)$ <u>is finitely generated as abelian group</u>.
(b) G <u>is accessible over</u> H.
(c) (<u>Terminal property</u>.) <u>Every nonempty family</u> $\mathcal{F}$ <u>of finitary decompositions of G over H has a terminal element, that is, such that no element of</u> $\mathcal{F}$ <u>is a proper expansion of it</u>.

Proof. (a) ⇒ (c). If (a) holds then we are in the situation of

5.11 and may, by the well-ordering, choose a decomposition D in $\mathcal{F}$ so as to minimize defect(D). Then by 5.11, D has no proper expansion belonging to $\mathcal{F}$.

(c) ⇒ (b). If (c) holds then, in particular, the (nonempty) family of <u>all</u> finitary decompositions of G over H has a terminal element, so G is accessible over H.

(b) ⇒ (a). If G is accessible over H then, by 5.6, $D_H^+(G, \mathbb{Z}[G])$ is finitely generated as left $\mathbb{Z}[G]$-module, so $D_H^+(G)$ is finitely generated as left $\mathbb{Z}$-module. □

Let us record an interesting application of 5.12.

5.13 COROLLARY. <u>If G is finitely generated over H and accessible over H then G is accessible over any subgroup K containing H.</u>

Proof. By 5.12 (b) ⇒ (c), the (nonempty) family of all finitary decompositions of G over K has a terminal element so G is accessible over K. □

5.14 REMARKS. By taking elementary proofs at every stage, we have bypassed some exact sequences that are normally used in this area. Since these are of interest in their own right, we now mention them briefly, but without much detail. Fix a decomposition $(G_v, q_e \mid v \in V, e \in E)$ of G over H, write Y for the underlying graph, and T for its maximal subtree. Set $E' = E(T)$, $E'' = E - E'$. For any right $R[G]$-module M, I.5.2 says that the right $\text{End}_{R[G]}(M)$-linear map

III   DECOMPOSITIONS

$$D_H^+(G,M) \longrightarrow M^{E''} \times \prod_V D_{H_v}^+(G_v,M)$$

$$d \mapsto ((q_e)d, (d|G_v))$$

is the kernel of a right $\text{End}_{R[G]}(M)$-linear map

(21) $\qquad M^{E''} \times \prod_V D_H^+(G_v,M) \longrightarrow \prod_E D_H^+(G_e,M)$

$$((m_e), (d_v)) \mapsto ((d_{1e} + ad\, m_e q_e^{-1} - (^{q_e})d_{\tau e} q_e^{-1}))$$

where we understand $m_e = 0$ for $e \in E'$. Observe that for any $v_0 \in V$, there is a right $\text{End}_{R[G]}(M)$-linear isomorphism

$$M^{E''} \times M^{V-\{v_0\}} \xrightarrow{\sim} M^E$$

$$((m_e), (m_v)) \mapsto ((m_{1e} + m_e q_e^{-1} - m_{\tau e} q_e^{-1})),$$

where again we understand $m_e = 0$ for $e \in E'$, and $m_{v_0} = 0$. Hence we have a commutative diagram

$$\begin{array}{ccccccccc}
0 & \to & \prod_{V-\{v_0\}} M^{G_v} & \to & M^{E''} \times M^{V-\{v_0\}} & \to & M^{E''} \times \prod_{V-\{v_0\}} \text{InD}(G_v,M) & \to & 0 \\
& & \downarrow & & \downarrow \wr & & \downarrow & & \\
0 & \to & \prod_E M^{G_e} & \to & M^E & \to & \prod_E \text{InD}(G_e,M) & \to & 0
\end{array}$$

and it follows that the vertical arrow which comes from (21) is surjective. (Notice that 4.12 now follows by the snake lemma, and the kernel of (12) can actually be computed. Notice that 5.5 also can be derived fairly easily from this fact.) This shows that $\prod_E \text{InD}(G_e,M)$ lies in the image of (21). So, for example, if $M = R[G]$ and all the G_e are finite then (21) is surjective. Thus if the decomposition is finitary we get an exact sequence

$$0 \to D_H^+(G, R[G]) \to R[G]^{E''} \oplus \bigoplus_V \bigoplus_{H_v} D_{H_v}^+(G_v, R[G]) \to \bigoplus_E R[G]/R[G]\sigma_{G_e} \to 0.$$

Taking $R = \mathbb{Z}$ and applying $\mathbb{Z} \otimes_{\mathbb{Z}[G]} -$ we get an exact sequence

$$\bigoplus_E \mathrm{Tor}_1^{\mathbb{Z}[G]}(\mathbb{Z}, \mathbb{Z}[G]/\mathbb{Z}[G]\sigma_{G_e}) \to D_H^+(G) \to \mathbb{Z}^{E''} \oplus \bigoplus_V \bigoplus_{H_v} D_{H_v}^+(G_v) \to \bigoplus_E D_{H_e}^+(G_e) \to 0.$$

Since $\omega \cap \mathbb{Z}[G]\sigma_{G_e} = \omega\sigma_{G_e}$, the Tor term vanishes, so we have a short exact sequence of abelian groups

$$0 \to D_H^+(G) \to \mathbb{Z}^{E''} \oplus \bigoplus_V \bigoplus_{H_v} D_{H_v}^+(G_v) \to \bigoplus_E D_{H_e}^+(G_e) \to 0. \quad \square$$

Our final topic of this section is the problem of uniqueness. The following result extends Lemma 7.6 of Scott-Wall [79].

5.15 THEOREM. <u>Suppose G is finitely generated over H and is accessible over H. Then there exists a reduced G-tree X with finite edge stabilizers, and having a vertex v_H stabilized by H, such that G_{v_H} is finitely generated over H and indecomposable over H, and for each $v \in V(X) - v_H G$, G_v is finitely generated and indecomposable.</u>

<u>For any such tree, X/G is finite, and $V(X)$, $E(X)$ are uniquely determined as G-sets, so X is determined as G-set.</u>

Proof. Since G is accessible over H, there exists a finitary decomposition of G over H such that each vertex group is finitely generated, and indecomposable, over its intersection with H. Let X be the corresponding right G-tree, as in 4.7. Then X/G is finite, so it can be contracted to a reduced G-tree, which then has all the desired properties.

Now let X' be any other such G-tree. Extend v_H to a

III   DECOMPOSITIONS

connected transversal V in V(X) for the G-action. For each $v \in V$, G_v is finitely generated over H_v, and indecomposable over H_v, so by 5.1, G_v stabilizes a vertex v of X'. This extends to a morphism of G-sets $\alpha: V(X) \to V(X')$; and by symmetry, there exists a morphism of G-sets $\beta: V(X') \to V(X)$. Now by 1.6, $V(X) \simeq V(X')$, $E(X) \simeq E(X')$ as G-sets. □

CHAPTER IV

COHOMOLOGICAL DIMENSION ONE

With his conjecture that every torsion-free group with a free subgroup of finite index is necessarily free, Serre essentially initiated the study of groups of cohomological dimension one. He showed that any such group has cohomological dimension at most one over $\mathbb{Z}$; subsequently, Stallings [68] showed that every finitely generated group of cohomological dimension at most one over $\mathbb{Z}$ is free, and Swan [69] then eliminated the finitely generated part of the hypothesis, thus proving the conjecture of Serre. Karrass-Pietrowski-Solitar [73], Cohen [73], and Scott [74] progressively extended the group theoretical part, concluding with the characterization of groups which have a free subgroup of finite index, as the fundamental groups of connected graphs of finite groups of bounded order. Returning to the cohomological aspect, Dunwoody [79] more generally characterized the groups which have cohomological dimension at most one over a given arbitrary nonzero ring, as the fundamental groups of connected graphs of finite groups having order invertible in the given ring. This deep result will now be obtained fairly painlessly from III.4.6. In fact, we prove a new, slightly stronger result, characterizing the transitive group actions that give rise to a projective augmentation module.

Throughout this chapter, let G be a group, H a subgroup of G, and R a nonzero ring.

IV COHOMOLOGICAL DIMENSION ONE

1. PROJECTIVE AUGMENTATION MODULES

Throughout this section, let U be a nonempty right G-set, and z be an element of U.

We are interested in determining when $\omega_R(H\backslash G)$ is $R[G]$-projective; this section considers, more generally, situations where $\omega_R(U)$ is $R[G]$-projective. At present, the known results are rather fragmentary, and we shall restrict ourselves to proving only what is needed later; for further information, see Wall [71].

We begin by looking at some <u>necessary</u> conditions for $\omega_R(U)$ to be $R[G]$-projective. The main idea is to apply III.4.6 to the G_z-derivation $G \to \omega_R(U)$, $g \mapsto z - zg$. We shall need some elementary facts, collected together in the next two lemmas.

We write R^{-1} for the set of those positive integers whose images in R are invertible in R; thus, for any set Y, we write $|Y| \in R^{-1}$ if Y is finite and its order is invertible in R.

1.1 LEMMA. (i) <u>If</u> M <u>is a projective right</u> $R[G]$-<u>module, then</u> M <u>is</u> $R[H]$-<u>projective</u>.
(ii) <u>If</u> N <u>is a right</u> $R[H]$-<u>module, then</u> N <u>is</u> $R[H]$-<u>projective if and only if</u> $N \otimes_{R[H]} R[G]$ <u>is</u> $R[G]$-<u>projective</u>.
(iii) <u>For any transversal</u> S <u>in</u> U <u>for the</u> G-<u>action,</u>
$R[U] \simeq \underset{s \in S}{\oplus} R \otimes_{R[G_s]} R[G]$ <u>as</u> $R[G]$-<u>module</u>.
(iv) $R[U]$ <u>is</u> $R[G]$-<u>projective if and only if</u> $|G_u| \in R^{-1}$ <u>for all</u> $u \in U$
(v) <u>If</u> G <u>fixes</u> z <u>then</u> $\omega_R(U) \simeq R[U-\{z\}]$, <u>and this is</u> $R[G]$-<u>projective if and only if</u> $|G_u| \in R^{-1}$ <u>for all</u> $u \in U-\{z\}$.
(vi) <u>If</u> $\omega_R(U)$ <u>is</u> $R[G]$-<u>projective, then for any distinct</u> x,y <u>in</u> U, $|G_{xy}| \in R^{-1}$, <u>where</u> $G_{xy} = G_x \cap G_y$.

Proof. (i) follows from the fact that $R[G] = \bigoplus_{G/H} R[gH]$ is free as right $R[H]$-module.

(ii). If N is $R[H]$-projective, then $N \otimes_{R[H]} R[G]$ is clearly $R[G]$-projective. Conversely, if $N \otimes_{R[H]} R[G]$ is $R[G]$-projective then it is $R[H]$-projective by (i), and hence, so are all the $R[H]$-summands in the decomposition $N \otimes_{R[H]} R[G] = \bigoplus_{H \backslash G/H} N \otimes_{R[H]} R[HgH]$, so in particular, N is $R[H]$-projective.

(iii) follows from III.4.11, since $R[U] \simeq \bigoplus_{s \in S} R[G_s \backslash G]$.

(iv). By (ii) and (iii), $R[U]$ is $R[G]$-projective if and only if R is $R[G_u]$-projective for all $u \in U$. Now observe that R is right $R[G]$-projective if and only if the augmentation map $\varepsilon_G : R[G] \to R$ has an $R[G]$-linear left inverse, which is equivalent to $R[G]^G$ containing an element p with $(p)\varepsilon_G = 1$. But $R[G]^G = R[\{\sigma_G\}]$, so $(R[G]^G)\varepsilon_G = Rn_G$ where $n_G = (\sigma_G)\varepsilon_G$. Thus R is $R[G]$-projective if and only if $|G| \in R^{-1}$, whence (iv).

(v). Suppose G fixes z. Then there is an $R[G]$-decomposition $R[U] = R[\{z\}] \oplus R[U-\{z\}]$, so the kernel of $R[U] \to R[\{z\}]$ is isomorphic to $R[U-\{z\}]$, but, by definition of $\omega_R(U)$, the kernel is also isomorphic to $\omega_R(U)$. This proves (v).

(vi). Suppose $\omega_R(U)$ is right $R[G]$-projective. Then for each $x \in U$, $\omega_R(U)$ is $R[G_x]$-projective by (i), and G_x fixes x, so by (v), $|G_{xy}| \in R^{-1}$ for all $y \in U-\{x\}$. □

For any family $\mathcal{F}$ of cardinals, we write $\underset{Y \in \mathcal{F}}{\text{HCF}}\, Y$ to denote the largest integer that divides every finite Y in $\mathcal{F}$, or 0 if $\mathcal{F}$ has no finite elements.

IV COHOMOLOGICAL DIMENSION ONE

1.2 LEMMA. *The following are equivalent.*

(a) *The G_z-derivation* $d_z: G \to \omega_R(U)$, $g \mapsto z - zg$, *is inner.*

(b) $0 \to \omega_R(U) \to R[U] \xrightarrow{\varepsilon_U} R \to 0$ *is* $R[G]$*-split.*

(c) $\underset{u \in U}{HCF} (G:G_u) \in R^{-1}$.

(d) $1 \in (R[U]^G)\varepsilon_U$.

If G_{xy} *is finite for all distinct* $x,y \in U$ *then these are further equivalent to*

(e) *Either* G *stabilizes an element of* U, *or* G *is finite and* $\underset{u \in U}{HCF} (G:G_u) \in R^{-1}$.

Proof. (a)⇔(d). We have the following equivalences:

d_z is inner

⇔ there exists $w \in \omega_R(U)$ such that $w - wg = z - zg$ for all $g \in G$

⇔ there exists $w \in \omega_R(U)$ such that $z - w \in R[U]^G$

⇔ $1 \in (R[U]^G)\varepsilon_U$.

(b)⇔(d) is clear.

(c)⇔(d). $R[U]^G = R[\{\sigma_{uG} \mid u \in U\}]$, so $(R[U]^G)\varepsilon_U = \Sigma R|uG|$, where the summation is over finite orbits uG. Since $\mathbb{Z}$ is a principal ideal domain, this latter ideal is principal, generated by the image of $\underset{u \in U}{HCF} (G:G_u)$ in R, which proves (c)⇔(d).

(e)⇒(c) is obvious.

Finally, if G_{xy} is finite for all distinct x,y in U then (c)⇒(e), for if G is infinite and (c) holds, then for some $u \in U$, $(G:G_u)$ is finite, so for any $g \in G$, $(G:G_u \cap G_{ug})$ is finite, so $G_u \cap G_{ug}$ is infinite, which means $u = ug$, so $g \in G_u$. Thus G fixes u, and (e) holds. □

PROJECTIVE AUGMENTATION IDEALS §1

1.3 THEOREM. *If* $\omega_R(U)$ *is* $R[G]$-*projective, and* G *is finitely generated over* G_z, *then there exists a finitary decomposition* $(G_v \mid v \in V)$ *of* G *over* G_z *such that for every* $v \in V$, *either* G_v *fixes an element of* U, *or* G_v *is finite and* $\mathrm{HCF}_{u \in U}(G_v : G_{vu}) \in R^{-1}$.

Proof. By 1.1 (vi), G_{xy} is finite for all distinct $x, y \in U$, so by III.4.6, and 1.2 (a) $\Rightarrow$ (e), the result follows. □

In the case $R = \mathbb{Z}$ the following result ensures that each G_v fixes an element of U.

1.4 PROPOSITION. *If* $G_{xy} = 1$ *for all distinct* $x, y \in U$, *and* $\mathrm{HCF}_{u \in U}(G : G_u) = 1$, *then* G *fixes an element of* U.

Proof. If G is infinite, this follows from 1.2, since here (c) $\Rightarrow$ (e). Thus we may assume that G is finite. For any $x, y \in U$ in different orbits, if $G_x \neq 1$ and $G_y \neq 1$ then

$$|G - \{1\}| \geq \left| \bigcup_{z \in xG \cup yG} (G_z - \{1\}) \right| = \sum_{z \in xG}(|G_z| - 1) + \sum_{z \in yG}(|G_z| - 1)$$

$= (G - (G:G_x)) + (G - (G:G_y)) \geq \tfrac{1}{2}|G| + \tfrac{1}{2}|G| = |G|$, which is impossible. It follows that there is an orbit xG such that G acts freely on $U - xG$, so $\mathrm{HCF}_{u \in U}(G : G_u) = (G : G_x)$, so $G = G_x$, and G fixes x. □

We now turn to <u>sufficient</u> conditions for $\omega_R(U)$ to be $R[G]$-projective. Here the key step is the following.

1.5 LEMMA. *If* M *is an* $R[G]$-*summand of* $R[U]$, *and* M *is* $R[G_u]$-*projective for all* $u \in U$, *then* M *is* $R[G]$-*projective*.

Proof. Choose a transversal S in U for the G-action, and for

IV COHOMOLOGICAL DIMENSION ONE

each $s \in S$, write m_s for the M-component of s within $R[U]$. Then m_s is fixed by G_s. By 1.1 (iii), there exists an $R[G]$-linear map $R[U] \to \bigoplus_{s \in S} M \otimes_{R[G_s]} R[G]$ sending each s to $m_s \otimes_{R[G_s]} 1$. The composite $R[U] \to \bigoplus_{s \in S} M \otimes_{R[G_s]} R[G] \to M$ sends each s to m_s, so is the given projection onto M. Thus M is isomorphic to an $R[G]$-summand of $\bigoplus_{s \in S} M \otimes_{R[G_s]} R[G]$, which is $R[G]$-projective by our hypotheses. Hence M is $R[G]$-projective. □

1.6 COROLLARY. <u>If</u> $\mathrm{HCF}(G : G_u) \in R^{-1}$, $u \in U$ <u>and</u> $|G_{xy}| \in R^{-1}$ <u>for all distinct</u> $x, y \in U$, <u>then</u> $\omega_R(U)$ <u>is</u> $R[G]$-<u>projective</u>.

Proof. By 1.2 (c) ⇒ (b), $\omega_R(U)$ is an $R[G]$-summand of $R[U]$, and by 1.1 (v), $\omega_R(U)$ is $R[G_u]$-projective for all $u \in U$, so by 1.5, $\omega_R(U)$ is $R[G]$-projective. □

We conclude this section with the most general conditions under which $\omega_R(U)$ is known to be $R[G]$-projective.

1.7 THEOREM. <u>Suppose that</u> $|G_{xy}| \in R^{-1}$ <u>for all distinct</u> x, y <u>in</u> U, <u>and suppose further that there exists a right</u> G-<u>tree</u> X <u>such that</u> U <u>embeds in</u> $V(X)$ <u>as</u> G-<u>set, and</u> $\mathrm{HCF}(G_v : G_{vu}) \in R^{-1}$ <u>for all</u> $v \in V(X)$. <u>Then</u> $\omega_R(U)$ <u>is</u> $R[G]$-<u>projective</u>.

Proof. Let V be a transversal in $V(X)$ for the G-action. By 1.2 (c) ⇒ (d), there is, for each $v \in V$, an element $x_v \in R[U]^{G_v}$ such that $(x_v)\varepsilon_U = 1$, and clearly, if $v \in U$, we may take $x_v = v$. This then extends to an $R[G]$-linear map $R[V(X)] \to R[U]$; moreover, this map is the identity on $R[U]$, and commutes with the augmentation maps. Thus, $\omega_R(U)$ is an $R[G]$-summand of $\omega_R(V(X))$.

By I.1.1, $\omega_R(V(X)) \simeq R[E(X)]$ as $R[G]$-modules, so by 1.5, it suffices to show that $\omega_R(U)$ is right $R[G_e]$-projective for all $e \in E(X)$. But it is clear from 1.6 that $\omega_R(U)$ is $R[G_v]$-projective for all $v \in V(X)$, which gives the desired result, by 1.1 (i), since $G_e \subseteq G_{\iota e}$ for all $e \in E(X)$. □

2. PAIRS OF GROUPS

If $\omega_R(G)$ is right $R[G]$-projective, one says that G has <u>cohomological dimension at most one</u> over R, and writes $cd_R G \leq 1$. We shall be concerned with the situation where $\omega_R(H\backslash G)$ is $R[G]$-projective, which includes $cd_R G \leq 1$ as the special case $H = 1$.

(For the general defintion of cohomological dimension cf Gruenberg [70], or Cohen [72]. We remark that the groups of cohomological dimension <u>precisely</u> one are the <u>infinite</u> groups of cohomological dimension at most one.)

Our main result chacterizes $\omega_R(H\backslash G)$ being $R[G]$-projective by the following concept.

2.1 DEFINITION. Let us say that H is an R^{-1}-<u>vertex</u> of G if there exists a decomposition $(G_v, q_e \mid v \in V(Y), e \in E(Y))$ of G over H, such that the following hold:

(1)   For each $e \in E(Y)$, G_e is finite;
(2)   For each $v \in V(Y)$, either G_v is finite, or $G_v = H$;
(3)   The following integers are invertible in R:

$\quad\quad\quad |H \cap H^g|$, $g \in G - H$;
$\quad\quad\quad \underset{g \in G}{HCF} (G_v : G_v \cap H^g)$, $v \in V(Y)$.

IV   COHOMOLOGICAL DIMENSION ONE

(We shall see that if H is an R^1-vertex of G then (3) is satisfied in <u>any</u> decomposition satisfying (1) and (2).) Notice that if H is infinite then $G_{v_H} = H$, so for all $g \in G - H$, $H \cap H^g = G_{v_H} \cap G_{v_H}^g$ is finite by II.3.1. We shall usually assume that $G_{v_H} = H$; notice that there is no loss of generality in this, for if H is finite then G_{v_H} is finite by (2), and we can adjoin a new edge e with new initial vertex, and having terminal vertex v_H, and we then set $G_{\iota e} = G_e = H$, and take ιe to be the new distinguished vertex.

We call a $\mathbb{Q}^1$-vertex of G a <u>vertex</u> of G; here the condition (3) is trivially satisfied, so this coincides with the definition of vertex given by Cohen [73]. □

Before going on to prove the equivalence of $\omega_R(H\backslash G)$ being $R[G]$-projective with H being an R^1-vertex of G, let us note the extreme case.

2.2 PROPOSITION. H <u>is a</u> $\mathbb{Z}^1$-<u>vertex of</u> G <u>if and only if</u> $G = H \amalg F$ <u>for a free group</u> F.

Proof. Suppose $G = H \amalg F$ for a free group F, say F is free on a set $\{q_e \mid e \in E\}$. Let Y be the graph with exactly one vertex, v, and with edge set E. Setting $G_v = H$ and $G_e = 1$ for all $e \in E$ defines a graph of groups whose fundamental group is isomorphic to G. Thus there is a decomposition $(G_v, q_e \mid v \in V(Y), e \in E(Y))$ of G over H with trivial edge groups. It follows that for all $g \in G - H$, $H \cap H^g = G_v \cap G_v^g = 1$ by II.3.1. It is now clear that H is a $\mathbb{Z}^1$-vertex of G.

Conversely, suppose H is a $\mathbb{Z}^1$-vertex of G, and let

$(G_v, q_e \mid v \in V(Y), e \in E(Y))$ be a decomposition as in 2.1, with $G_{v_H} = H$. So $H \cap H^g = 1$ for all $g \in G - H$, and, by 1.4, for each $v \in V(Y)$, $G_v \subseteq H^g$ for some $g \in G$. Thus, if $G_v \neq 1$, there is a unique conjugate of H containing G_v. Let X be the standard right G-tree, as in III.4.7. Consider the set $\{v \in V(X) \mid 1 \neq G_v \subseteq H\}$. This is an H-subset of $V(X)$; moreover, it is the vertex set of a subtree of X, and two elements are in the same H-orbit if and only if they are in the same G-orbit in X. Thus, we may choose a connected transversal for the H-action in the subtree, and extend it to a connected transversal in X for the G-action. This gives a (new) decomposition of G over H for which $G_v \subseteq H$ for all $v \in V(Y)$. It follows that we may assume that Y has exactly one vertex, and that the decomposition is of the form $(H, q_e \mid e \in E(Y))$. Now since $H \cap H^{q_e} = 1$ for all $e \in E(Y)$, we see from this presentation that $G = H \amalg F$, where F is the free group freely generated by the q_e. □

Returning to the characterization of pairs (G,H) such that $\omega_R(H \backslash G)$ is $R[G]$-projective, we see that sufficiency is immediate from 1.7.

2.3 THEOREM. *If* H *is an* R^{-1}-*vertex of* G *then* $\omega_R(H \backslash G)$ *is* $R[G]$-*projective*. □

We now proceed to show the converse, starting by reducing the problem with the following observation.

2.4 LEMMA. *If* $\omega_R(H \backslash G)$ *is* $R[G]$-*projective then* H *is an* R^{-1}-*vertex of* G *if and only if it is a vertex of* G.

IV   COHOMOLOGICAL DIMENSION ONE

Proof. One direction is trivial. In the opposite direction, we see from 1.1 (vi) that $|H \cap H^g| \in R^{-1}$ for all $g \in G - H$, and it remains to consider a decomposition $(G_v, q_e \mid v \in V(Y), e \in E(Y))$ of G over H satisfying (1) and (2), and show that the second part of (3) holds. Let $v \in V(Y)$. By 1.1 (i), $\omega_R(H\backslash G)$ is $R[G_v]$-projective, so the derivation $G_v \to \omega_R(H\backslash G)$, $g \mapsto H(1-g)$, is inner. (For, we need consider only the case where G_v is finite, by (2), and here any derivation from G_v to a projective $R[G_v]$-module is inner by III.4.2.) So, by 1.2 (a) $\Rightarrow$ (c), $\operatorname*{HCF}_{g \in G}(G_v : G_v \cap H^g) \in R^{-1}$. Thus H is an R^{-1}-vertex of G. $\square$

The problem that remains is to show that if $\omega_R(H\backslash G)$ is $R[G]$-projective, then H is a vertex of G. We begin with a result that comes from the finitely generated case, and is suitable for getting information in the general case.

2.5 LEMMA. <u>Suppose</u> $H \leq K \leq L \leq G$, <u>and</u> L <u>is finitely generated over</u> H. <u>If</u> K <u>is the kernel of a derivation from</u> G <u>to a projective</u> $R[G]$-<u>module, then</u> K <u>is finitely generated over</u> H.

<u>If moreover</u>, $\omega_R(H\backslash G)$ <u>is</u> $R[G]$-<u>projective, and</u> L <u>also is the kernel of a derivation from</u> G <u>to a projective</u> $R[G]$-<u>module, then</u> K <u>is a vertex of</u> L.

Proof. If K is the kernel of a derivation from G to a projective $R[G]$-module, then by 1.1 (i), K is the kernel of a derivation from L to a projective $R[L]$-module. So, by III.4.6, there exists a finitary decomposition $(L_w \mid w \in W)$ of L, having K as one of the vertex groups, say L_{w_K}. By III.3.3, K is finitely generated over H.

Now suppose further that $\omega_R(H\backslash G)$ is $R[G]$-projective, hence $R[L]$-projective, and consider the derivation $L \to \omega_R(H\backslash G)$. By 1.3, there exists an L-tree X with finite edge stabilizers, such that H fixes a vertex v_H of X, and for each vertex v of X, either L_v is finite or $L_v \leq H^g$ for some $g \in G$. To prove that K is a vertex of L, it suffices to show that for each $w \in W - \{w_K\}$, the action of L_w on the L-tree X has all vertex stabilizers finite; for then, each such L_w can be expanded by III.1.2 to be replaced with finite groups, which would show that K is indeed a vertex of L.

Thus, let $w \in W - \{w_K\}$, and let L_w act on X. For each vertex v of X, the stabilizer of v is either finite or contained in H^g for some $g \in G$, so it suffices to show that $L \cap H^g$ is finite for all $g \in G$. If $g \in L$, then $L_w \cap H^g \subseteq L_w \cap K^g = L_w \cap L^g_{w_K}$, and this is finite by II.3.1. This leaves the case where $g \in G - L$. Since $L_w \cap H^g \subseteq L \cap L^g$, it suffices to show that for $g \in G - L$, $L \cap L^g$ is finite. For this, we invoke the hypothesis that L is the kernel of a derivation d from G to a projective $R[G]$-module. Now for any $x \in L \cap L^g$, say $gx = yg$, with $x,y \in L$, we have $(gx)d = (yg)d$, so $(g)d.x + (x)d = (y)d.g + (g)d$, so $(g)d.x = (g)d$ since $x,y \in L = \ker d$. This says that $L \cap L^g$ fixes the nonzero element $(g)d$ of a projective $R[G]$-module, and the only way for this to happen is for $L \cap L^g$ to be finite, as desired. □

IV   COHOMOLOGICAL DIMENSION ONE

(We remark that the foregoing proves substantially more than is claimed, but in its present form, 2.5 is ideal for our purposes.)

The method for extending from 2.5 to the general case is based on arguments of Cohen [73], which were in turn based on arguments of Swan [69]. We shall need two more lemmas just to get to the countably generated case.

2.6 LEMMA (Cohen [73]). <u>Let γ be an ordinal, and suppose $(G_\beta)_{1 \leq \beta \leq \gamma}$ is an ascending chain of subgroups of G, such that for each $\alpha < \gamma$, G_α is a vertex of $G_{\alpha+1}$, and for each limit ordinal $\beta \leq \gamma$, $G_\beta = \underset{\alpha < \beta}{\cup} G_\alpha$. Then G_1 is a vertex of G_γ.</u>

Proof. We shall construct, by transfinite induction, an ascending chain $(T_\beta)_{1 \leq \beta \leq \gamma}$ of trees, such that, for each $\beta \leq \gamma$,

T_β is a right G_β-tree, and there is a distinguished vertex v_1 with stabilizer G_1, and all other orbits have finite stabilizers;

if $\beta = \alpha+1$ there is an embedding $T_\alpha \otimes_{G_\alpha} G_\beta \to T_\beta$ of G_β-graphs, that preserves the distinguished vertex, $v_1 \otimes 1 \to v_1$;

if β is a limit ordinal, then $T_\beta = \underset{\alpha < \beta}{\cup} T_\alpha$.

Define $T_1 = \{v_1\}$ with trivial G_1-action.

Suppose $\beta \leq \gamma$ and we have defined T_α for all $\alpha < \beta$.

Consider first the case where $\beta = \alpha+1$. Since G_α is a vertex of G_β, there is a G_β-tree $X = X(\alpha)$, such that one vertex v_α has stabilizer G_α, and all other orbits have finite stabilizers. Now G_α acts on T_α, so by III.1.1, we can expand v_α in X to

PAIRS OF GROUPS §2

get a new G_β-tree, T_β, having

$$V(T_\beta) = (V(X) - v_\alpha G_\beta) \vee (V(T_\alpha) \otimes_{G_\alpha} G_\beta),$$

$$E(T_\beta) = E(X) \vee (E(T_\alpha) \otimes_{G_\alpha} G_\beta),$$

and there is a distinguished vertex, $v_1 \otimes 1$, with stabilizer G_1, such that all other orbits have finite stabilizers, and there is an embedding $T_\alpha \otimes_{G_\alpha} G_\beta \to T_\beta$ of G_β-graphs.

If β is a limit ordinal, we define $T_\beta = \bigcup_{\alpha < \beta} T_\alpha$. Since $G_\beta = \bigcup_{\alpha < \beta} G_\alpha$, T_β is a G_β-tree, and since stabilizers are preserved at each step in the chain, T_β has all the desired properties.

Hence G_1 is a vertex of G_β for all $\beta \le \gamma$. In particular, G_1 is a vertex of G_γ. □

For $H \le K \le G$, we write $\omega_R(H \backslash K)G$ for the $R[G]$-submodule of $\omega_R(H \backslash G)$ generated by the natural image of $\omega_R(H \backslash K)$. For convenience, we summarise some simple observations.

2.7 LEMMA. <u>Let</u> $H \le K \le L \le G$.

(i) $\omega_R(H \backslash K)G \simeq \omega_R(H \backslash K) \otimes_{R[K]} R[G]$.

(ii) $\omega_R(H \backslash G) \simeq \omega_R(G)/\omega_R(H)G$.

(iii) $\omega_R(H \backslash K)G \simeq \omega_R(K)G/\omega_R(H)G$.

(iv) $\omega_R(K \backslash L)G \simeq \omega_R(H \backslash L)G/\omega_R(H \backslash K)G$.

(v) <u>For any subset</u> S <u>of</u> G, <u>if</u> $S \cup H$ <u>generates</u> K, <u>then</u> $\{H(1-s) \mid s \in S\}$ <u>generates</u> $\omega_R(H \backslash K)G$ <u>as</u> $R[G]$-<u>module</u>.

(vi) <u>If</u> $\omega_R(H \backslash K)G \supseteq \omega_R(H \backslash K_1)$ <u>for some</u> $H \le K_1 \le G$ <u>then</u> $K \supseteq K_1$.

(vii) <u>If</u> S <u>is a subset of</u> G <u>such that</u> $\omega_R(H \backslash K)G$ <u>is generated as</u> $R[G]$-<u>module by</u> $\{H(1-s) \mid s \in S\}$ <u>then</u> $H \cup S$ <u>generates</u> K.

IV COHOMOLOGICAL DIMENSION ONE

Proof. (i) follows from the fact that $R[G] = \bigoplus_{K\backslash G} R[Kg]$ is flat as left $R[K]$-module.

(ii). $R \otimes_{R[H]} R[G] = R[H\backslash G]$ by III.4.11, and this induces an isomorphism $R[G]/\omega_R(H)G \simeq R[H\backslash G]$, and (ii) follows.

(iii). $\omega_R(H\backslash K)G \simeq \omega_R(H\backslash K) \otimes_{R[K]} R[G] \simeq (\omega_R(K)/\omega_R(H)K) \otimes_{R[K]} R[G]$.

(iv) is clear from (iii).

(v). Let M be the $R[G]$-submodule of $\omega_R(H\backslash G)$ generated by $\{H(1-s) \mid s \in S\}$. Then $M \subseteq \omega_R(H\backslash K)G$, and conversely, the derivation $G \to \omega_R(H\backslash G)/M$, $g \mapsto H(1-g) + M$, vanishes on $H \cup S$, so vanishes on K, so $\omega_R(H\backslash K)G \subseteq M$.

(vi). For any $g \in K_1$, $H(1-g) \in \omega_R(H\backslash K)G$, so $K(1-g) = 0$ in $\omega_R(K\backslash G)$. Thus $g \in K$, so $K_1 \subseteq K$.

(vii). Let K_1 be the subgroup generated by $S \cup H$. Then by (v), $\omega_R(H\backslash K_1)G = \omega_R(H\backslash K)G$, so by (vi), $K = K_1$. □

We are now in a position to prove the key step.

2.8 THE COUNTABLY GENERATED CASE. *If $\omega_R(H\backslash G)$ is countably generated and projective as $R[G]$-module, then H is a vertex of G*

Proof. Since $\omega_R(H\backslash G)$ is generated by the elements of the form $H(1-g)$, $g \in G$, we may choose a generating set $y_1, y_2, \ldots$ where for each n, $y_n = H(1-g_n)$ for some $g_n \in G$.

Let F be a right $R[G]$-module that is free on a set indexed by the positive integers, so there is a natural surjection $F \to \omega_R(H\backslash G)$. Let $\phi = (\phi_n)$ be a left inverse of this map; that is, for each positive integer n, $\phi_n : \omega_R(H\backslash G) \to R[G]$ is a functional, and for each $w \in \omega_R(H\backslash G)$, $(w)\phi_n = 0$ for almost all n, and $w = \sum_{n\geq 1} y_n \cdot (w)\phi_n$. (Thus we are choosing a "projective coordinate system" for $\omega_R(H\backslash G)$.) For each $n \geq 1$, we have a derivation $d_n: G \to R[G]$, $g \mapsto (H(1-g))\phi_n$, and for each $g \in G$, $H(1-g) = \sum_{n\geq 1} y_n \cdot (g) d_n$.

For each positive integer m, let $G_m = \{g \in G \mid gd_n = 0 \text{ for all } n \geq m\}$. Notice that G_m is the kernel of a derivation from G to a free $R[G]$-module, $g \mapsto (gd_n)_{n \geq m}$. Also $H = G_1 \leq G_2 \leq \ldots$, and $\bigcup_{m \geq 1} G_m = G$.

For each positive integer m, let L_m be the subgroup of G generated by $H \cup \{g_1, \ldots, g_m\}$, so $\{y_1, \ldots, y_m\} \subseteq \omega_R(H\backslash L_m)$. Thus $\omega_R(H\backslash G_m) \subseteq \omega_R(H\backslash L_m)G$, so by 2.7 (vi), $G_m \leq L_m$. Hence by the first part of 2.5, G_m is finitely generated over H. So of course G_{m+1} is finitely generated over H. Now by the second part of 2.5, G_m is a vertex of G_{m+1}, and then by 2.6, $H = G_1$ is a vertex of G, as claimed. □

To pass to the general case we rely on the following classical result, slightly modified for our purposes.

2.9 THEOREM (Kaplansky [58]). <u>Let</u> P <u>be a projective right</u> R-<u>module, and</u> Y <u>any generating set for</u> P. <u>Then there exists an ordinal</u> γ, <u>and an ascending chain</u> $(P_\beta)_{1 \leq \beta \leq \gamma}$ <u>of submodules of</u> P <u>such that</u> $P_1 = 0$, $P_\gamma = P$, <u>and for each</u> $\beta \leq \gamma$, P_β <u>is generated by a subset of</u> Y, <u>and if</u> β <u>is a limit ordinal then</u> $P_\beta = \bigcup_{\alpha < \beta} P_\alpha$, <u>and if</u> $\beta = \alpha+1$ <u>then</u> P_β/P_α <u>is a countably generated projective</u> R-<u>module</u>.

Proof. As in the preceding proof, there exist functionals $\phi_y: P \to R$, $y \in Y$, such that for each $p \in P$, $(p)\phi_y = 0$ for almost all $y \in Y$, and $p = \sum_{y \in Y} y \cdot (p) \phi_y$.

Let $\mathcal{S}$ be the set of subsets A of Y such that for all $a \in A$, $y \in Y - A$, we have $(a)\phi_y = 0$. Clearly $\mathcal{S}$ is closed under arbitrary unions. For any subset A of Y, we construct an element $\bar{A}$ of $\mathcal{S}$ as follows. Define $A_0 = A$, and inductively define

$$A_{n+1} = \{y \in Y \mid (a)\phi_y \neq 0 \text{ for some } a \in A_n\}.$$

Now set $\bar{A} = \bigcup_{n \geq 0} A_n$. It is fairly easy to see that $\bar{A}$ belongs to $\mathcal{S}$, and is the smallest element of $\mathcal{S}$ containing A. Notice that if A is finite then so is each A_n, so $\bar{A}$ is countable.

We inductively construct an ascending chain in $\mathcal{S}$ as follows. We set $Y_1 = \phi$. Now suppose β is an ordinal and we have constructed Y_α for all $\alpha < \beta$. If $\beta = \alpha+1$, then either $Y_\alpha R = P$, and we stop, or there exists some element $p \in P - Y_\alpha R$, and we define $Y_\beta = Y_\alpha \cup \overline{\{p\}}$. If β is a limit ordinal, then we define $Y_\beta = \bigcup_{\alpha < \beta} Y_\alpha$. Since Y is a set, this process eventually gives an ordinal γ such that $Y_\gamma R = P$.

For each $\beta \leq \gamma$, we set $P_\beta = Y_\beta R$. We claim that the resulting chain of submodules of P has all the desired properties. The only property that is not immediate is that for each $\alpha < \gamma$, $P_{\alpha+1}/P_\alpha$ is projective. Observe that P_α is a direct summand of P, with retraction $P \to P_\alpha$, $p \mapsto \sum_{y \in Y_\alpha} y \cdot (p)\phi_y$. Hence each P_α is projective, and P_α is a summand of $P_{\alpha+1}$, which completes the proof. □

We now arrive at the main result. With the additional hypothesis that $cd_R G \leq 1$, this was proved by Dunwoody [79], building on work of Stallings, Swan, Cohen et al.

PAIRS OF GROUPS §2

2.10 THEOREM. $\omega_R(H\backslash G)$ is $R[G]$-projective if and only if H is an R^{-1}-vertex of G.

Proof. From 2.3 and 2.4, it remains to show that if $\omega_R(H\backslash G)$ is $R[G]$-projective then H is a vertex of G. Consider the generating set $Y = \{H(1-g) \mid g \in G\}$, and let $(P_\beta)_{1 \leq \beta \leq \gamma}$ be a chain of $R[G]$-submodules of $\omega_R(H\backslash G)$ as in 2.9. By 2.7 (v), we have, for each $\beta \leq \gamma$, $P_\beta = \omega_R(H\backslash G_\beta)G$ for some subgroup G_β of G containing H. By 2.7 (vi), we have a chain $(G_\beta)_{1 \leq \beta \leq \gamma}$ of subgroups of G, such that $G_1 = H$, $G_\gamma = G$, and for each $\beta \leq \gamma$, if β is a limit ordinal then $G_\beta = \bigcup_{\alpha < \beta} G_\alpha$, and if $\beta = \alpha+1$, then

$$P_\beta/P_\alpha = (\omega_R(H\backslash G_\beta)G)/(\omega_R(H\backslash G_\alpha)G)$$
$$\simeq \omega_R(G_\alpha\backslash G_\beta)G \text{ by 2.7 (iv),}$$
$$\simeq \omega_R(G_\alpha\backslash G_\beta) \otimes_{R[G_\beta]} R[G] \text{ by 2.7 (i).}$$

From 2.9, we are given that this is a countably generated projective $R[G]$-module; by 1.1 (ii), $\omega_R(G_\alpha\backslash G_\beta)$ is then a projective $R[G_\beta]$-module, and it is easily seen that it must be countably generated. So by 2.8, G_α is a vertex of G_β, so by 2.6, H is a vertex of G. □

In the case $R = \mathbb{Z}$, 2.10 and 2.3 give an equivalence that was conjecture by Wall [71], and proved by Dunwoody [79] in the case where G is finitely generated over H.

2.11 THEOREM. $\omega_\mathbb{Z}(H\backslash G)$ is $\mathbb{Z}[G]$-projective if and only if $G = H \amalg F$ for a free group F. Moreover, in this event, if F is freely generated by $\{q_e \mid e \in E\}$ then $\omega_R(H\backslash G)$ is $R[G]$-free on $\{H(1-q_e) \mid e \in E\}$.

IV   COHOMOLOGICAL DIMENSION ONE

Proof. The second part follows from the fact that G acts on a tree X having $V(X) = H\backslash G$, $E(X) = \bigvee_E 1\backslash G$, and by I.1.1, $\omega_R(V(X)) \simeq R[E(X)]$ as $R[G]$-modules. □

We now return to 2.10 and record the case $H = 1$.

2.12 THEOREM (Dunwoody [79]). *The following are equivalent.*

(a) $\omega_R(G)$ *is right* $R[G]$-*projective.*
(b) G *is the fundamental group of a connected graph of finite groups whose orders are invertible in* R.
(c) G *acts on a tree* X *in such a way that the orders of the stabilizers of the vertices are finite and invertible in* R. □

3.   FINITE EXTENSIONS OF FREE GROUPS

The implication of 2.12 for abstract group theory is the promised converse of II.3.6. We first require a result; let us call G an R^{-1}-*group* if all its finite subgroups have order invertible in R.

3.1 LEMMA. *If* $(G:H)$ *is finite, and* G *is an* R^{-1}-*group, and* $cd_R H \le 1$, *then* $cd_R G \le 1$.

Proof. Clearly $R[H] + \omega_R(G) = R[G]$, and $R[H] \cap \omega_R(G) = \omega_R(H)$, so we have an exact sequence $0 \to \omega_R(H) \to R[H] \oplus \omega_R(G) \to R[G] \to 0$ of right $R[H]$-modules. Since $R[G]$ is right $R[H]$-projective, this sequence splits, so if $\omega_R(H)$ is $R[H]$-projective, then $\omega_R(G)$ is $R[H]$-projective.

Now consider first the case where $(G:H) \in R^{-1}$. Here, for any right $R[G]$-module M, the multiplication map $M \otimes_{R[H]} R[G] \to M$ has

FINITE EXTENSIONS OF FREE GROUPS §3

a left inverse $M \to M \otimes_{R[H]} R[G]$, $m \mapsto (G:H)^{-1} \sum_{Hg \in H\backslash G} mg^{-1} \otimes g$, so if M is $R[H]$-projective, then M is $R[G]$-projective. Thus $\omega_R(G)$ is right $R[G]$-projective in the case where $(G:H) \in R^{-1}$.

In the general case, we see from 2.12 that $cd_{\mathbb{Q}} H \leq 1$ so, by the preceding, $cd_{\mathbb{Q}} G \leq 1$, so by 2.12 again, $cd_R G \leq 1$ since G is an R^{-1}-group. □

Originally, Serre proved the analogue of 3.1 for <u>arbitrary</u> cohomological dimension, in the case where R is <u>commutative</u>, cf Cohen [72], p.9. In the case $R = \mathbb{Z}$, Serre [71], §1.7, gives a topological proof. It would be interesting to know if Serre's argument could be refined to show that if $(G:H)$ is finite and H acts on a tree with, say, finite edge stabilizers, then G acts on some tree with the stabilizers being finite extensions of the H-stabilizers. Such a construction would illuminate the next result, the converse of II.3.6. The finitely generated case of this result is due to Karrass-Pietrowski-Solitar [73], the countably generated case to Cohen [73], and the final step to Scott [74].

3.2 THEOREM. <u>G has a free subgroup of finite index if and only if G is the fundamental group of a connected (faithful) graph of finite groups of bounded order.</u>

Proof. One direction is II.3.6. Conversely, suppose G has a free subgroup F of finite index. Then every finite subgroup of G acts freely on G/F so has order dividing $(G:F)$. Further, $cd_{\mathbb{Q}} F \leq 1$ by 2.11 or 2.12, so by 3.1, $cd_{\mathbb{Q}} G \leq 1$, so by 2.12, G is the fundamental group of a connected faithful graph of finite groups, whose order is bounded by $(G:F)$. □

IV   COHOMOLOGICAL DIMENSION ONE

Let us note some consequences of 2.12 and 3.2.

3.3 THEOREM. The following are equivalent.
(a) $cd_R G \leq 1$, and the finite subgroups of G are of bounded order.
(b) G is an R^{-1}-group, and has a free subgroup of finite index.
(c) G is the fundamental group of a connected (faithful) graph of finite R^{-1}-groups of bounded order.

Proof. (a)⇔(c) by 2.12, and (b)⇔(c) by 3.2. □

3.4 THEOREM. The following are equivalent.
(a) $cd_R G \leq 1$, and G is finitely generated.
(b) G is an R^{-1}-group, and has a finitely generated free subgroup of finite index.
(c) G is the fundamental group of a finite connected (faithful) graph of finite R^{-1}-groups.

Proof. (a)⇔(c) by 2.12, and (b)⇔(c) by 3.2. □

It is appropriate that we should conclude with the $R = \mathbb{Z}$ case of 3.3.

3.5 THEOREM (Serre-Stallings-Swan). The following are equivalent.
(a) $cd_{\mathbb{Z}} G \leq 1$.
(b) G is torsion-free, and has a free subgroup of finite index.
(c) G is free.

Proof. (a)⇔(c) by 2.12, and (b)⇔(c) by 3.2. □

BIBLIOGRAPHY AND AUTHOR INDEX

The page references at the end of the entries indicate the places in the text where the entry is quoted; other references to an author are listed after his name.

Bamford, C. and Dunwoody, M.J.
76. On accessible groups. *J. Pure and Applied Algebra* **7** (1976) 333-346. [88,90

Bass, H. [vi,20,21,23,35,36,37,39,41,48,55
76. Some remarks on groups acting on trees. *Comm. Algebra* **4** (1976) 1091-1126. [27

Bergman, G.M.
68. On groups acting on locally finite graphs. *Ann. of Math.* **88** (1968) 335-340. [66

Britton, J.L.
63. The word problem. *Ann.of Math.* **77** (1963) 16-32. [44

Chiswell, I.M. [vi,vii,21
73. *On groups acting on trees*. PhD Thesis, University of Michigan, Ann Arbor, Michigan, 1973. [20,21
76. Exact sequences associated with a graph of groups. *J. Pure and Applied Algebra* **8** (1976) 63-74. [20
76'. The Grushko-Neumann Theorem. *Proc. London Mat. Soc.(3)* **33** (1976) 385-400. [50
77. *Groups acting on trees*. Lecture course at Queen Mary College, London, 1976/7. [21,50
79. The Bass-Serre Theorem revisited. *J. Pure and Applied Algebra* **15** (1979) 117-123. [21

Cohen, D.E. [vii,116
70. Ends and free products of groups. *Math. Zeit.* **114** (1970) 9-18. [63,66
72. *Groups of cohomological dimension one*. Lecture Notes in Mathematics, Vol. 245. Springer, Berlin, 1972. [vi,107,119
73. Groups with free subgroups of finite index. pp26-44 in *Conference on Group Theory, University of Wisconsin-Parkside 1972*. Lecture Notes in Mathematics, Vol. 319. Springer, 1973. [58,101,108,112,119

Dicks, W.
77. Mayer-Vietoris presentations over colimits of rings. *Proc. London Math. Soc.(3)* **34** (1977) 557-576. [21
79. Hereditary group rings. *J. London Math. Soc.* **20** (1979) 27-39. [21

Dunwoody, M.J. *see also* Bamford, C. [v,vi,87
79. Accessibility and groups of cohomological dimension one. *Proc. London Math. Soc.(3)* **38** (1979) 193-215. [27,31,34,55,66,69,72,77,82,88,90 92,96,101,116,117,118

BIBLIOGRAPHY AND AUTHOR INDEX

Gruenberg, K.W.

70. *Cohomological topics in group theory.* Lecture Notes in Mathematics, Vol. 143. Springer, Berlin, 1970. [107

Grushko, I.A.

40. Über die Basen eines freien Produktes von Gruppen. *Mat.Sb.* 8 (1940) 169-182. (Russian. German summary.) [53

Higgins, P.J. [vi

66. Grushko's Theorem. *J.Algebra* 4 (1966) 365-372. [49

Higman, G., Neumann, B.H. and Neumann, H.

49. Embedding theorems for groups. *J.London Math.Soc.* 24 (1949) 247-254. [14

Kaplansky, I.

58. Projective modules. *Ann.of Math.* 68 (1958) 372-377. [115

Karrass, A., Pietrowski, A. and Solitar, D.

73. Finite and infinite cyclic extensions of free groups. *J.Australian Math.Soc.* 16 (1973) 458-466. [46,101,119

Kurosh, A.G.

37. Zum Zerlegungsproblem der Theorie der freien Produkte. *Rec.Math.Moscow* 2 (1937) 995-1001. (Russian. German summary.) [48

Lyndon, R.C. and Schupp, P.E.

77. *Combinatorial group theory.* Ergebnisse der Mathematik 89. Springer, Berlin, 1977.

Neumann, B.H. *see also* **Higman, G.**

43. On the number of generators of a free product. *J.London Math.Soc.* 18 (1943) 12-20. [53

Neumann, H. *see also* **Higman, G.**

48. Generalised free products with amalgamated subgroups I. *Amer.J.Math.* 70 (1948) 590-625. [48

Passman, D.S.

77. *The algebraic structure of group rings.* John Wiley and Sons, London 1977.

Pietrowski, A. *see* **Karrass, A.**

Reidemeister, K.

32. *Einführung in die kombinatorische Topologie.* Vieweg, Braunschweig, 1932. (Reprinted by Chelsea, New York, 1950.) [36

Schupp, P.E. *see* **Lyndon, R.C.**

Scott, G.P.

74. An embedding theorem for groups with a free subgroup of finite index. *Bull.London Math.Soc.* 6 (1974) 304-306. [101,119

BIBLIOGRAPHY AND AUTHOR INDEX

Scott, G.P. and Wall, C.T.C.
79. Topological methods in group theory, pp137-203 in *Homological Group Theory*. LMS Lecture Notes 36. Cambridge University Press, Cambridge, 1979. [vi,79

Schreier, O.
27. Die Untergruppen der freien Gruppen. *Abh. Math. Univ. Hamburg* **5** (1927) 161-183. [36

Serre, J.-P. [v,vi,35,36,55,101,119,120
71. Cohomologie des groupes discrets, pp77-169 in *Prospects in Mathematics*. Ann. of Math. Studies 70. Princeton, 1971. [119
77. *Arbres, amalgames et SL_2*. Astérisque No. 46. Société Math.de France, 1977. [12,20,21,23,24,27,37,39,41,47,48

Solitar, D. *see* **Karrass, A.**

Stallings, J.R. [v,vi,116,120
68. On torsion-free groups with infinitely many ends. *Ann. of Math.* **88** (1968) 312-334. [55,85,101

Swan, R.G. [v,116,120
69. Groups of cohomological dimension one. *J. Algebra* **12** (1969) 585-610. [101,112

Wagner, D.H.
57. On free products of groups. *Trans. Amer. Math. Soc.* **84** (1957) 352-378. [52

Wall, C.T.C. *see also* **Scott, G.P.**
71. Pairs of relative cohomological dimension one. *J. Pure and Applied Algebra* **1** (1971) 141-154. [87,102,117

SUBJECT INDEX

Accessible (over a subgroup), 87
Act, 7, 8
Act freely, 36
Admit (a derivation), 74
Almost contain, 30, 63
Almost equal, 30, 63
Almost-right-invariant, 30, 63, 75
Almost right subset, 30, 69
Augmentation ideal, 99
Augmentation map, 79
Augmentation module, 79
Automorphism of graphs, 4

Bouquet of loops, 36

Cayley graph, 8
Chosen successors, 67
Circuit, 2
Coboundary, 62
Cohomological dimension (one), 107
Colimit, 48
Component, 2
Connected, 2, 6, 10, 63
Connecting family, 9
Contract, 61
Coproduct, 48
Coproduct with amalgamation, 14
Cover, 32
Covering, 6
Cut, 62

Decomposable (over a subgroup), 85
Decomposition, 55
Decomposition over a subgroup, 69
Defect, 95
Derivation, 16

Edge, 1
Edge group, 10, 56
Expand, 73, 58
Expansion, 95
Extremity, 27

Faithful, 41
Fibre bundle, 58
Finitary, 69
Finitely generated over a subgroup, 68
Fix = stabilize, 27
Forest, 2
Free action, 36
Free product, 48
Free product with amalgamation, 14

Full, 66
Fundamental group, 10, 40

G-action, 7
G-bimodule, 16
G-graph, 8
G-set, 7, 8
G-tree, 8, 79
Geodesic, 2
Generate finitely over a subgroup, 68
Graph, 1
Graph automorphism, 4
Graph of groups, 10
Group action, 7
Group ring, 15
Groups, 10

H-derivation, 75
HNN construction, 14
HNN extension, 15

Improper expansion, 96
Inaccessible (over a subgroup), 87
Indecomposable (over a subgroup), 85
Induced G-set, 56
Infinite path, see path
Initial, 1
Initial extremity, 27
Inner derivation, 16
Interval, 31
Isomorphism of graphs, 4

Length, 2
Local isomorphism, 5
Locally injective, 5
Locally surjective, 5
Loop, 1

Maximal subtree, 3, 6
Morphism of graphs, 4

Outer derivation, 82
Orbit, 7

Path, 1
Path format, 44
Path in universal specialization, 40
Points to, points away from, 29
Proper decomposition, 82
Proper expansion, 96
Proper S-cut, 63
Proper almost-right-invariant set, 82

SUBJECT INDEX

R^{-1}-group, 118
R^{-1}-vertex, 107
Rank, 52
Reduced G-tree, 61
Reduced path, 2, 44
Reduced form, 2
Reduction, 2, 44
Right G-set, 8
Right G-tree, 79

S-coboundary, 62
S-cut, 63
Shift, 28
Simple reduction, 2
Size, 94
Source, 29
Spanning tree, 3
Specialization, 38
Stabilizer, 7
Standard graph, 12, 40
Standard tree, 12, 40, 79

Star, 4
Structure Theorem, 23
Subgraph, 1

Terminal, 1
Terminal extremity, 27
Thin, 66
Transversal, 6, 7
Tree, 2
Tree of groups, 48
Tree product, 48
Trivial, 35, 56

Underlying path, 41
Unreduced G-tree, 61
Universal covering, 6
Universal specialization, 39

Vertex, 1
Vertex group, 10, 55
Vertex of a group, 108

SYMBOL INDEX

$\underline{a},\overline{a}$,30,63

ad,16,75

$cd_R G$,107

$\mathbb{C}$,viii

$D(\ ,\ )$,74

$D_*(\ ,\ )$,75

$D_*^+(\ ,\ )$,75

$D_*^+(\ )$,90

$\mathfrak{D}(R[[\ ]])$,75

$\mathfrak{D}_*(R[[\ ]])$,76

$\mathfrak{D}_*^+(R[[\ ]])$,76

δ,62

δ_G,75

δ_S,62

E,74

$E(\ )$,1,4,59,60

e^1,e^{-1},1

ε,91

ε_U,79

$(\ )^G$,60,75

G_x,7

G_y,43

$G(y)$,21

G_e^ε,43

$G[e]$,30

$G[e,v]$,29,82

Groups,10

$\Gamma(\ ,\ )$,8,12,41

H,24

H_v,71

HCF,103

HNN,14

ι,1

ι_e,10

$InD(\ ,\ )$,75

$\pi(\ ,\ )$,10,35,40

$\pi(\ )$,48

π_y,15

$P(G)$,64

$\overline{P}(G)$,65

$P_S(G)$,65

Π,viii

$\mathbb{Q}$,viii

q_e,9

R^{-1},102,109,118

$R[\ ]$,3,15

$R[[\ ]]$,75

ρ,24

S^1,24

SL_2,24

Sym,7

$\mathcal{S}$,66,116

σ_*,75

τ,1

τ_e,10

t,38,94

$U(\ )$,39

$V(\ )$,1,4

$V\delta$,62

v_H,69

ω,99

$\omega_R(\ )$,79

$\omega_R(\ )G$,113

$X(\ ,\ )$,3

χ_*,76

$\mathbb{Z}$,viii

*,32,63

$*_v$,73

[ ],64

[,],16,32

< >,13,30

~,32

≤,≥,27,65,94,95

$\underset{S}{\leq}$,65

≼,64

\,/,7,8,49

Ⅱ,14,48

^,15,57,76

v,∨,viii

| |,viii

-,viii

φ,viii

⊕,viii

⊗,viii

QA
3
L28
v.790

JUN 26 1980

Vol. 640: J. L. Dupont, Curvature and Characteristic Classes. X, 175 pages. 1978.

Vol. 641: Séminaire d'Algèbre Paul Dubreil, Proceedings Paris 1976–1977. Edité par M. P. Malliavin. IV, 367 pages. 1978.

Vol. 642: Theory and Applications of Graphs, Proceedings, Michigan 1976. Edited by Y. Alavi and D. R. Lick. XIV, 635 pages. 1978.

Vol. 643: M. Davis, Multiaxial Actions on Manifolds. VI, 141 pages. 1978.

Vol. 644: Vector Space Measures and Applications I, Proceedings 1977. Edited by R. M. Aron and S. Dineen. VIII, 451 pages. 1978.

Vol. 645: Vector Space Measures and Applications II, Proceedings 1977. Edited by R. M. Aron and S. Dineen. VIII, 218 pages. 1978.

Vol. 646: O. Tammi, Extremum Problems for Bounded Univalent Functions. VIII, 313 pages. 1978.

Vol. 647: L. J. Ratliff, Jr., Chain Conjectures in Ring Theory. VIII, 133 pages. 1978.

Vol. 648: Nonlinear Partial Differential Equations and Applications, Proceedings, Indiana 1976–1977. Edited by J. M. Chadam. VI, 206 pages. 1978.

Vol. 649: Séminaire de Probabilités XII, Proceedings, Strasbourg, 1976–1977. Edité par C. Dellacherie, P. A. Meyer et M. Weil. VIII, 805 pages. 1978.

Vol. 650: C*-Algebras and Applications to Physics. Proceedings 1977. Edited by H. Araki and R. V. Kadison. V, 192 pages. 1978.

Vol. 651: P. W. Michor, Functors and Categories of Banach Spaces. VI, 99 pages. 1978.

Vol. 652: Differential Topology, Foliations and Gelfand-Fuks-Cohomology, Proceedings 1976. Edited by P. A. Schweitzer. XIV, 252 pages. 1978.

Vol. 653: Locally Interacting Systems and Their Application in Biology. Proceedings, 1976. Edited by R. L. Dobrushin, V. I. Kryukov and A. L. Toom. XI, 202 pages. 1978.

Vol. 654: J. P. Buhler, Icosahedral Golois Representations. III, 143 pages. 1978.

Vol. 655: R. Baeza, Quadratic Forms Over Semilocal Rings. VI, 199 pages. 1978.

Vol. 656: Probability Theory on Vector Spaces. Proceedings, 1977. Edited by A. Weron. VIII, 274 pages. 1978.

Vol. 657: Geometric Applications of Homotopy Theory I, Proceedings 1977. Edited by M. G. Barratt and M. E. Mahowald. VIII, 459 pages. 1978.

Vol. 658: Geometric Applications of Homotopy Theory II, Proceedings 1977. Edited by M. G. Barratt and M. E. Mahowald. VIII, 487 pages. 1978.

Vol. 659: Bruckner, Differentiation of Real Functions. X, 247 pages. 1978.

Vol. 660: Equations aux Dérivée Partielles. Proceedings, 1977. Edité par Pham The Lai. VI, 216 pages. 1978.

Vol. 661: P. T. Johnstone, R. Paré, R. D. Rosebrugh, D. Schumacher, R. J. Wood, and G. C. Wraith, Indexed Categories and Their Applications. VII, 260 pages. 1978.

Vol. 662: Akin, The Metric Theory of Banach Manifolds. XIX, 306 pages. 1978.

Vol. 663: J. F. Berglund, H. D. Junghenn, P. Milnes, Compact Right Topological Semigroups and Generalizations of Almost Periodicity. X, 243 pages. 1978.

Vol. 664: Algebraic and Geometric Topology, Proceedings, 1977. Edited by K. C. Millett. XI, 240 pages. 1978.

Vol. 665: Journées d'Analyse Non Linéaire. Proceedings, 1977. Edité par P. Bénilan et J. Robert. VIII, 256 pages. 1978.

Vol. 666: B. Beauzamy, Espaces d'Interpolation Réels: Topologie et Géometrie. X, 104 pages. 1978.

Vol. 667: J. Gilewicz, Approximants de Padé. XIV, 511 pages. 1978.

Vol. 668: The Structure of Attractors in Dynamical Systems. Proceedings, 1977. Edited by J. C. Martin, N. G. Markley and W. Perrizo. VI, 264 pages. 1978.

Vol. 669: Higher Set Theory. Proceedings, 1977. Edited by G. H. Müller and D. S. Scott. XII, 476 pages. 1978.

Vol. 670: Fonctions de Plusieurs Variables Complexes III, Proceedings, 1977. Edité par F. Norguet. XII, 394 pages. 1978.

Vol. 671: R. T. Smythe and J. C. Wierman, First-Passage Perculation on the Square Lattice. VIII, 196 pages. 1978.

Vol. 672: R. L. Taylor, Stochastic Convergence of Weighted Sums of Random Elements in Linear Spaces. VII, 216 pages. 1978.

Vol. 673: Algebraic Topology, Proceedings 1977. Edited by P. Hoffman, R. Piccinini and D. Sjerve. VI, 278 pages. 1978.

Vol. 674: Z. Fiedorowicz and S. Priddy, Homology of Classical Groups Over Finite Fields and Their Associated Infinite Loop Spaces. VI, 434 pages. 1978.

Vol. 675: J. Galambos and S. Kotz, Characterizations of Probability Distributions. VIII, 169 pages. 1978.

Vol. 676: Differential Geometrical Methods in Mathematical Physics II, Proceedings, 1977. Edited by K. Bleuler, H. R. Petry and A. Reetz. VI, 626 pages. 1978.

Vol. 677: Séminaire Bourbaki, vol. 1976/77, Exposés 489–506. IV, 264 pages. 1978.

Vol. 678: D. Dacunha-Castelle, H. Heyer et B. Roynette. Ecole d'Eté de Probabilités de Saint-Flour. VII-1977. Edité par P. L. Hennequin. IX, 379 pages. 1978.

Vol. 679: Numerical Treatment of Differential Equations in Applications, Proceedings, 1977. Edited by R. Ansorge and W. Törnig. IX, 163 pages. 1978.

Vol. 680: Mathematical Control Theory, Proceedings, 1977. Edited by W. A. Coppel. IX, 257 pages. 1978.

Vol. 681: Séminaire de Théorie du Potentiel Paris, No. 3, Directeurs: M. Brelot, G. Choquet et J. Deny. Rédacteurs: F. Hirsch et G. Mokobodzki. VII, 294 pages. 1978.

Vol. 682: G. D. James, The Representation Theory of the Symmetric Groups. V, 156 pages. 1978.

Vol. 683: Variétés Analytiques Compactes, Proceedings, 1977. Edité par Y. Hervier et A. Hirschowitz. V, 248 pages. 1978.

Vol. 684: E. E. Rosinger, Distributions and Nonlinear Partial Differential Equations. XI, 146 pages. 1978.

Vol. 685: Knot Theory, Proceedings, 1977. Edited by J. C. Hausmann. VII, 311 pages. 1978.

Vol. 686: Combinatorial Mathematics, Proceedings, 1977. Edited by D. A. Holton and J. Seberry. IX, 353 pages. 1978.

Vol. 687: Algebraic Geometry, Proceedings, 1977. Edited by L. D. Olson. V, 244 pages. 1978.

Vol. 688: J. Dydak and J. Segal, Shape Theory. VI, 150 pages. 1978.

Vol. 689: Cabal Seminar 76-77, Proceedings, 1976-77. Edited by A.S. Kechris and Y. N. Moschovakis. V, 282 pages. 1978.

Vol. 690: W. J. J. Rey, Robust Statistical Methods. VI, 128 pages. 1978.

Vol. 691: G. Viennot, Algèbres de Lie Libres et Monoïdes Libres. III, 124 pages. 1978.

Vol. 692: T. Husain and S. M. Khaleelulla, Barrelledness in Topological and Ordered Vector Spaces. IX, 258 pages. 1978.

Vol. 693: Hilbert Space Operators, Proceedings, 1977. Edited by J. M. Bachar Jr. and D. W. Hadwin. VIII, 184 pages. 1978.

Vol. 694: Séminaire Pierre Lelong – Henri Skoda (Analyse) Année 1976/77. VII, 334 pages. 1978.

Vol. 695: Measure Theory Applications to Stochastic Analysis, Proceedings, 1977. Edited by G. Kallianpur and D. Kölzow. XII, 261 pages. 1978.

Vol. 696: P. J. Feinsilver, Special Functions, Probability Semigroups, and Hamiltonian Flows. VI, 112 pages. 1978.

Vol. 697: Topics in Algebra, Proceedings, 1978. Edited by M. F. Newman. XI, 229 pages. 1978.

Vol. 698: E. Grosswald, Bessel Polynomials. XIV, 182 pages. 1978.

Vol. 699: R. E. Greene and H.-H. Wu, Function Theory on Manifolds Which Possess a Pole. III, 215 pages. 1979.

Vol. 700: Module Theory, Proceedings, 1977. Edited by C. Faith and S. Wiegand. X, 239 pages. 1979.

Vol. 701: Functional Analysis Methods in Numerical Analysis, Proceedings, 1977. Edited by M. Zuhair Nashed. VII, 333 pages. 1979.

Vol. 702: Yuri N. Bibikov, Local Theory of Nonlinear Analytic Ordinary Differential Equations. IX, 147 pages. 1979.

Vol. 703: Equadiff IV, Proceedings, 1977. Edited by J. Fábera. XIX, 441 pages. 1979.

Vol. 704: Computing Methods in Applied Sciences and Engineering, 1977, I. Proceedings, 1977. Edited by R. Glowinski and J. L. Lions. VI, 391 pages. 1979.

Vol. 705: O. Forster und K. Knorr, Konstruktion verseller Familien kompakter komplexer Räume. VII, 141 Seiten. 1979.

Vol. 706: Probability Measures on Groups, Proceedings, 1978. Edited by H. Heyer. XIII, 348 pages. 1979.

Vol. 707: R. Zielke, Discontinuous Čebyšev Systems. VI, 111 pages. 1979.

Vol. 708: J. P. Jouanolou, Equations de Pfaff algébriques. V, 255 pages. 1979.

Vol. 709: Probability in Banach Spaces II. Proceedings, 1978. Edited by A. Beck. V, 205 pages. 1979.

Vol. 710: Séminaire Bourbaki vol. 1977/78, Exposés 507-524. IV, 328 pages. 1979.

Vol. 711: Asymptotic Analysis. Edited by F. Verhulst. V, 240 pages. 1979.

Vol. 712: Equations Différentielles et Systèmes de Pfaff dans le Champ Complexe. Edité par R. Gérard et J.-P. Ramis. V, 364 pages. 1979.

Vol. 713: Séminaire de Théorie du Potentiel, Paris No. 4. Edité par F. Hirsch et G. Mokobodzki. VII, 281 pages. 1979.

Vol. 714: J. Jacod, Calcul Stochastique et Problèmes de Martingales. X, 539 pages. 1979.

Vol. 715: Inder Bir S. Passi, Group Rings and Their Augmentation Ideals. VI, 137 pages. 1979.

Vol. 716: M. A. Scheunert, The Theory of Lie Superalgebras. X, 271 pages. 1979.

Vol. 717: Grosser, Bidualräume und Vervollständigungen von Banachmoduln. III, 209 pages. 1979.

Vol. 718: J. Ferrante and C. W. Rackoff, The Computational Complexity of Logical Theories. X, 243 pages. 1979.

Vol. 719: Categorial Topology, Proceedings, 1978. Edited by H. Herrlich and G. Preuß. XII, 420 pages. 1979.

Vol. 720: E. Dubinsky, The Structure of Nuclear Fréchet Spaces. V, 187 pages. 1979.

Vol. 721: Séminaire de Probabilités XIII. Proceedings, Strasbourg, 1977/78. Edité par C. Dellacherie, P. A. Meyer et M. Weil. VII, 647 pages. 1979.

Vol. 722: Topology of Low-Dimensional Manifolds. Proceedings, 1977. Edited by R. Fenn. VI, 154 pages. 1979.

Vol. 723: W. Brandal, Commutative Rings whose Finitely Generated Modules Decompose. II, 116 pages. 1979.

Vol. 724: D. Griffeath, Additive and Cancellative Interacting Particle Systems. V, 108 pages. 1979.

Vol. 725: Algèbres d'Opérateurs. Proceedings, 1978. Edité par P. de la Harpe. VII, 309 pages. 1979.

Vol. 726: Y.-C. Wong, Schwartz Spaces, Nuclear Spaces and Tensor Products. VI, 418 pages. 1979.

Vol. 727: Y. Saito, Spectral Representations for Schrödinger Operators With Long-Range Potentials. V, 149 pages. 1979.

Vol. 728: Non-Commutative Harmonic Analysis. Proceedings, 1978. Edited by J. Carmona and M. Vergne. V, 244 pages. 1979.

Vol. 729: Ergodic Theory. Proceedings, 1978. Edited by M. Denker and K. Jacobs. XII, 209 pages. 1979.

Vol. 730: Functional Differential Equations and Approximation of Fixed Points. Proceedings, 1978. Edited by H.-O. Peitgen and H.-O. Walther. XV, 503 pages. 1979.

Vol. 731: Y. Nakagami and M. Takesaki, Duality for Crossed Products of von Neumann Algebras. IX, 139 pages. 1979.

Vol. 732: Algebraic Geometry. Proceedings, 1978. Edited by K. Lønsted. IV, 658 pages. 1979.

Vol. 733: F. Bloom, Modern Differential Geometric Techniques in the Theory of Continuous Distributions of Dislocations. XII, 206 pages. 1979.

Vol. 734: Ring Theory, Waterloo, 1978. Proceedings, 1978. Edited by D. Handelman and J. Lawrence. XI, 352 pages. 1979.

Vol. 735: B. Aupetit, Propriétés Spectrales des Algèbres de Banach. XII, 192 pages. 1979.

Vol. 736: E. Behrends, M-Structure and the Banach-Stone Theorem. X, 217 pages. 1979.

Vol. 737: Volterra Equations. Proceedings 1978. Edited by S.-O. Londen and O. J. Staffans. VIII, 314 pages. 1979.

Vol. 738: P. E. Conner, Differentiable Periodic Maps. 2nd edition, IV, 181 pages. 1979.

Vol. 739: Analyse Harmonique sur les Groupes de Lie II. Proceedings, 1976-78. Edited by P. Eymard et al. VI, 646 pages. 1979.

Vol. 740: Séminaire d'Algèbre Paul Dubreil. Proceedings, 1977-78. Edited by M.-P. Malliavin. V, 456 pages. 1979.

Vol. 741: Algebraic Topology, Waterloo 1978. Proceedings. Edited by P. Hoffman and V. Snaith. XI, 655 pages. 1979.

Vol. 742: K. Clancey, Seminormal Operators. VII, 125 pages. 1979.

Vol. 743: Romanian-Finnish Seminar on Complex Analysis. Proceedings, 1976. Edited by C. Andreian Cazacu et al. XVI, 713 pages. 1979.

Vol. 744: I. Reiner and K. W. Roggenkamp, Integral Representations. VIII, 275 pages. 1979.

Vol. 745: D. K. Haley, Equational Compactness in Rings. III, 167 pages. 1979.

Vol. 746: P. Hoffman, τ-Rings and Wreath Product Representations. V, 148 pages. 1979.

Vol. 747: Complex Analysis, Joensuu 1978. Proceedings, 1978. Edited by I. Laine, O. Lehto and T. Sorvali. XV, 450 pages. 1979.

Vol. 748: Combinatorial Mathematics VI. Proceedings, 1978. Edited by A. F. Horadam and W. D. Wallis. IX, 206 pages. 1979.

Vol. 749: V. Girault and P.-A. Raviart, Finite Element Approximation of the Navier-Stokes Equations. VII, 200 pages. 1979.

Vol. 750: J. C. Jantzen, Moduln mit einem höchsten Gewicht. III, 195 Seiten. 1979.

Vol. 751: Number Theory, Carbondale 1979. Proceedings. Edited by M. B. Nathanson. V, 342 pages. 1979.

Vol. 752: M. Barr, *-Autonomous Categories. VI, 140 pages. 1979.

Vol. 753: Applications of Sheaves. Proceedings, 1977. Edited by M. Fourman, C. Mulvey and D. Scott. XIV, 779 pages. 1979.

Vol. 754: O. A. Laudal, Formal Moduli of Algebraic Structures. III, 161 pages. 1979.

Vol. 755: Global Analysis. Proceedings, 1978. Edited by M. Grmela and J. E. Marsden. VII, 377 pages. 1979.

Vol. 756: H. O. Cordes, Elliptic Pseudo-Differential Operators – An Abstract Theory. IX, 331 pages. 1979.

Vol. 757: Smoothing Techniques for Curve Estimation. Proceedings, 1979. Edited by Th. Gasser and M. Rosenblatt. V, 245 pages. 1979.

Vol. 758: C. Năstăsescu and F. Van Oystaeyen; Graded and Filtered Rings and Modules. X, 148 pages. 1979.